MILITARY AVIATION IN THE GULF SOUTH

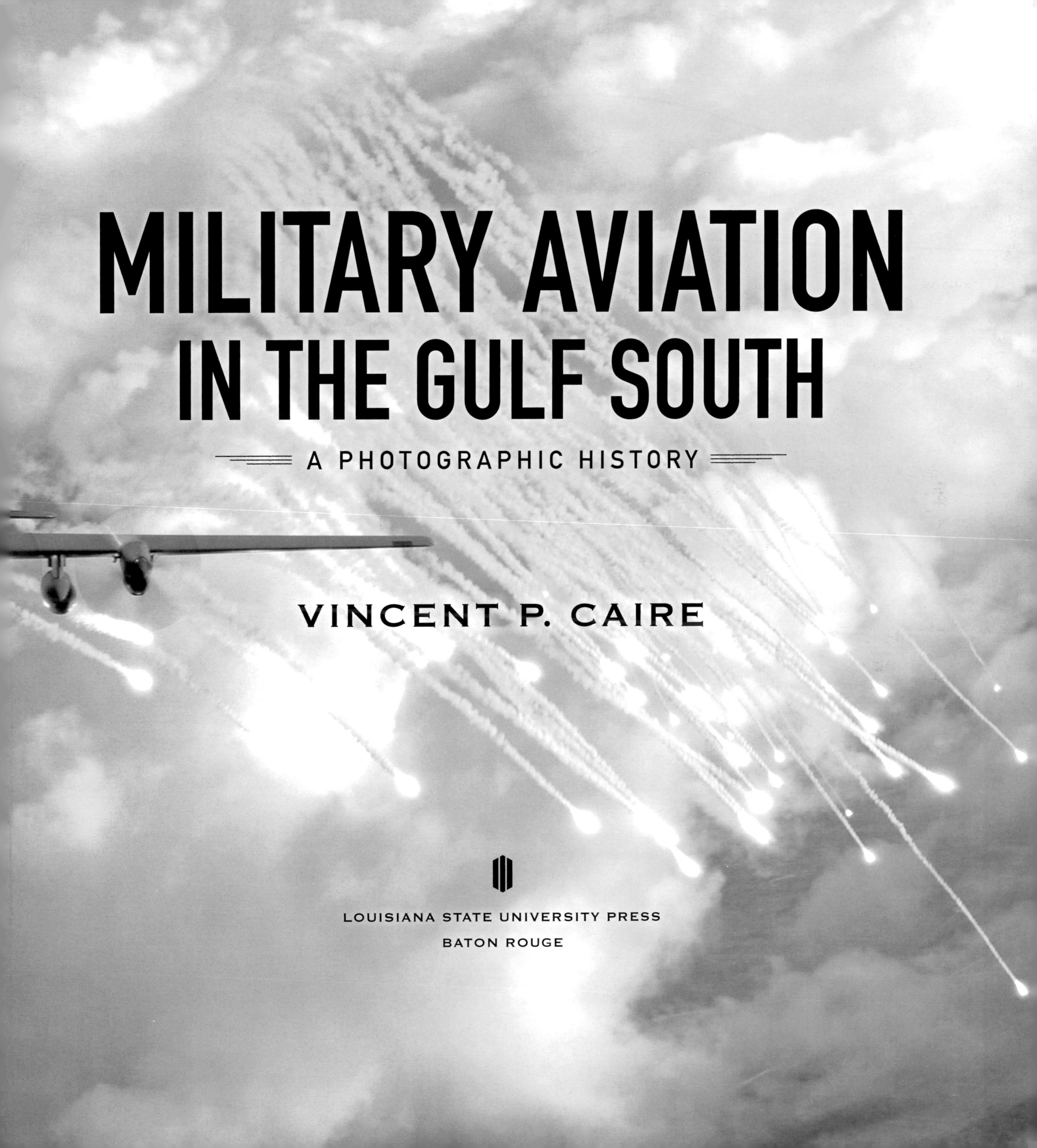

MILITARY AVIATION IN THE GULF SOUTH

A PHOTOGRAPHIC HISTORY

VINCENT P. CAIRE

LOUISIANA STATE UNIVERSITY PRESS
BATON ROUGE

Published by Louisiana State University Press

Manufactured in the United States of America
First printing

DESIGNER: Barbara Neely Bourgoyne
TYPEFACE: Whitman
PRINTER AND BINDER: Thomson-Shore, Inc.

Photo opposite: An F-35 Lightning II Joint Strike Fighter at Eglin Air Force Base, Florida. (U.S. Air Force photo/Samuel King Jr.)

The views expressed by the author, and other opinions in this book, do not necessarily reflect the official views or policies of the U.S. Department of Defense, the U.S. Air Force, the U.S. Army, the U.S. Coast Guard, the U.S. Marine Corps, or the U.S. Navy.

LIBRARY OF CONGRESS CATALOGING-IN-PUBLICATION DATA
Names: Caire, Vincent P., 1962–
Title: Military aviation in the Gulf South : a photographic history / Vincent P. Caire.
Description: Baton Rouge : Louisiana State University Press, [2016] | Includes index.
Identifiers: LCCN 2016008112| ISBN 978-0-8071-6411-2 (cloth : alk. paper) | ISBN 978-0-8071-6412-9 (pdf) | ISBN 978-0-8071-6413-6 (epub) | ISBN 978-0-8071-6414-3 (mobi)
Subjects: LCSH: Aeronautics, Military—United States—Pictorial works. | Aeronautics, Military—Southern States—Pictorial works. | Air bases—United States—Pictorial works. | Air bases—Southern States—Pictorial works. | Airplanes, Military—United States—History—Pictorial works. | Southern States—History, Military—Pictorial works.
Classification: LCC UG633 .C1493 2016 | DDC 358.400976—dc23
LC record available at http://lccn.loc.gov/2016008112

The paper in this book meets the guidelines for permanence and durability of the Committee on Production Guidelines for Book Longevity of the Council on Library Resources. ♾

"ETERNAL VIGILANCE IS THE PRICE OF LIBERTY."

These well-known, energizing words enshrine our cherished freedoms.

Never let that vigilance, provided by the men, women, and families of the armed services of the United States of America, be underappreciated.

Your sacrifices, past, present, and future, far exceed any capacity for repayment.

THANK YOU!

CONTENTS

1910. Orville Wright (*third from left, standing*), accompanied by students, prepares a Wright Flyer for takeoff on the grounds of the Frank Kohn cotton plantation. Today, the property is the site of Maxwell Air Force Base. (U.S. Air Force photo)

Left: **1923.** The USS *Langley* (CV-1), the first aircraft carrier operated by the U.S. Navy, is resupplied in port at NAS Pensacola. Affectionately referred to as "the Covered Wagon," the *Langley* was commissioned as a carrier in 1922 and the following year conducted training exercises in the Gulf of Mexico and Caribbean waters. In February 1942, the *Langley* was attacked and mortally damaged by Japanese aircraft in the Pacific theater near Java and later sunk by its own escort vessels to prevent it from falling into enemy hands. (Photo courtesy of the Public Affairs Office, NAS Pensacola)

A NOTE ON MILITARY AVIATION IN THE GULF COAST REGION

Powered flight dawned in the United States along the shores of the Atlantic Ocean when the Wright Flyer took the sky over Kitty Hawk, North Carolina. Yet it was a stretch of land bordering another body of water, the Gulf of Mexico, that played a key role in taking Orville and Wilbur Wright's invention to greater heights in defense of the nation. In fact, the brothers chose land that would one day be home to Maxwell Air Force Base outside Montgomery, Alabama, as the site for a flying field to demonstrate their new invention. It was only the beginning.

In January 1914, the U.S. Navy's entire aviation establishment, consisting of a handful of personnel and varied assortment of aircraft, arrived in Pensacola, Florida, to establish the sea service's first air station there. Just months later, aviators from Pensacola would deploy to the waters off Veracruz, Mexico, where they would log the first combat flights in U.S. military history. While the "Cradle of Naval Aviation" was the first significant military air base operational on the Gulf Coast, it would certainly not be the only one. The portrait of military aviation in this region so ably presented by Vincent Caire in his pictorial history captures this chapter of the nation's flying heritage.

With its temperate, relatively predictable climate and wide-open, unpopulated spaces that existed before the post–World War II population boom of the Sun Belt, the southern states bordering the Gulf of Mexico proved ideal for air operations, particularly the training of generations of military pilots wearing uniforms representing all the U.S. military services. Army Air Corps (later U.S. Air Force) pilots, among them Charles Lindbergh, flew from fields around San Antonio, Texas, in pursuit of their silver wings, while a Louisiana native named Claire Chennault commanded a pursuit squadron at Barksdale Field in his home state before leaving the U.S. Army to head to the Far East. There he would command the American Volunteer Group, more famously known as the Flying Tigers.

The global war in which the famous shark-mouth P-40s of the Flying Tigers flew into action spawned a wide expansion of military aviation activities on the Gulf Coast. Eglin Field in Fort Walton Beach, which in 1942 was the training site of the famed Doolittle Raiders for their heroic mission against Japan launched from the aircraft carrier *Hornet* (CV-8), was the scene of weapons testing throughout World War II. Houma, Louisiana, operated U.S. Coast Guard patrol planes and U.S. Navy airships that patrolled for German U-boats in the Gulf of Mexico, while Army Air Forces gunnery students, among them actor Clark Gable, passed through Tyndall Field in Panama City, Florida.

As the postwar years brought the jet age and the space race, the Gulf Coast region continued the mission of training military airmen, some of them eventually joining NASA and blasting off for epic voyages into outer space. The region also deployed squadrons of fighters and strategic bombers to Vietnam, the Middle East, and other far-flung battle sites. On September 11, 2001, Air Force One, carrying President George W. Bush, landed briefly at Barksdale Air Force Base before returning to Washington, D.C., the events of that day triggering the Global War on Terror—the latest chapter in the Gulf Coast's enduring contribution to our nation's history that is the subject of this outstanding work.

—Hill Goodspeed, Historian
National Naval Aviation Museum
Naval Air Station Pensacola, Florida

FOREWORD

When Vincent Caire contacted me to obtain historical photographs of Maxwell Air Force Base, Alabama, for a book that he was writing about the history of military aviation in the Gulf South, coincidentally I had recently given a presentation at a regional history conference on the untapped wealth of information in U.S. Air Force history offices in the Gulf South states.

In researching that presentation, I discovered that the Air Force and its predecessors have had a significant presence in the Gulf states from the early years of military aviation in the United States in the 1910s to the present. The height of that presence was during World War II, when the Army Air Forces had almost 70 main airfields and several hundred auxiliary fields in the Gulf South states. Although the Air Force has inactivated all but 16 of those main airfields and most of the auxiliary fields since 1946, their influence in the region is still felt; about 94 percent of them were converted to civil airports by nearby towns, cities, and county governments. Thus, the Air Force continues to have a major impact on the history, economy, society, and culture of the Gulf South states.

In addition to the Air Force's presence in the Gulf South, Caire also covers the extensive presence of the U.S. Navy, the U.S. Marine Corps, and the U.S. Army aviation installations in the region. As a result, his book demonstrates the significant influence and impact that military aviation has had on the Gulf South, a part of history that many may not know.

The book's many photographs bring the narrative alive. Overall, it is a welcome addition to the historiography of the Gulf South, telling the long story of military aviation in this section of the United States.

—Robert B. Kane
Director of History, Air University
Maxwell Air Force Base, Alabama

PBY Catalina flying boats on the flight line during World War II at Naval Air Station Jacksonville, Florida. (Photo courtesy of the Emil Buehler Library, National Naval Aviation Museum, NAS Pensacola, FL)

PREFACE

I would like to begin by clarifying what this book is, and is not. This work presents a century (plus a few years) of military aviation history in the Gulf South; it is a celebration in story. It is not an attempt to present an encyclopedic, all-inclusive chronicle. I believe, especially when it comes to the subject of military history, that publishing comprehensive works, while lending the appearance of being complete, is an impossible mission. These pages do not feature a listing of every squadron temporarily or permanently assigned to the Gulf South over the past one hundred years, nor do they pretend to reference every event in military aviation that has occurred in the region. Such a collection of data would not support my intention; rather, it would be an extremely large work of multiple volumes destined for the library reference stacks.

Had I embarked upon that comprehensive route, someone or something—be it those who served in short-lived, less familiar squadrons, or one of the multitude of secret operations that preserved democracy, unfortunately now known only to their former personnel or the deepest archive collections—would be unintentionally overlooked. That is not the end result I had in mind for this book.

My objective for this work is simple. Through a broad introduction, it is my desire to energize the reader, to encourage appreciation or even further exploration of featured subjects after getting to know about them here, possibly for the first time. For instance, few outside of the military appreciate that the majority of basic pilot training in every branch of our services occurs right here in the Gulf South. One of the purposes of this book is to recognize and celebrate this regional partnership.

The challenge faced in writing a book about military aviation is that, in fact, execution of the defense mission is shared by bases throughout the country and not so conveniently divided along state lines. Each facility and region of our nation could easily support a book telling its own story. It is unfortunate that several have yet to be the subject of such narratives.

Although this book concentrates on the historical development of military aviation in the Gulf South, my purpose is not to elevate the naval air stations, Army airfields, and Air Force bases along this southern coastline above their counterparts in the rest of the United States. Rather, it is to share a sampling of the many histories and narratives associated with their mission, one made strong in no small measure because of the Gulf South's population, culture, and geography.

The majority of photographs included here, especially the contemporary images, were taken by U.S. military photographers. I am in their debt. Wherever possible, credit is given to the individual military photographer, whose original art-quality image becomes public property at the snap of the lens. These talented photographers seldom receive personal recognition outside of formal military publications.

Finally, it is appropriate for Louisiana State University Press to publish a book that honors the evolution of military aviation, specifically the course of that evolution in the Gulf South. I am proud to be the Press's partner in this project. LSU is a university with a deep history of preparing military officers and personnel for the duties of national defense; its press is one of the oldest academic publishers in the South.

Fondly referred to as the "Ole War Skule," LSU stands among other agricultural and mechanical colleges, state universities, and private institutions that over the last century have educated thousands of men and women who served in the conflicts this nation has endured. If that military service was in aviation, those personnel most certainly were assigned to at least one of the bases highlighted within these pages, and they are a part of this story.

MILITARY AVIATION IN THE GULF SOUTH

2013. An F-35C Lightning II aircraft piloted by Lt. Cmdr. Chris Tabert, assigned to Strike Fighter Squadron VFA-101, flies the squadron's first local sortie. The F-35C is the carrier variant of the Joint Strike Fighter. VFA-101, based at Eglin Air Force Base, Florida, serves as the F-35C fleet replacement squadron, training both air crew and maintenance personnel to fly and repair the F-35C. (U.S. Air Force/Samuel King Jr.)

INTRODUCTION

For over one hundred years, military aviation has had an active presence in the Gulf South. Spanning Texas to Florida, this region hosted the first formal flight training programs for every branch of the U.S. armed services. It has been home to military bases with specialized missions in historic aviation operations, ranging from the 1916 pursuit of Pancho Villa into Mexico, to the famous Doolittle Raid over Tokyo in April 1942, to operational development of the "bunker bombs" and smart weapons technologies used in the twenty-first-century Middle East wars.

The history runs deep. Many contemporary Air Force bases and naval air stations were founded on the footprints of Army forts and Navy yards past. From the early 1800s—when the Gulf Coast served as the back door to the budding United States of America—through today, the presence and activity of the military have reinforced coastal communities located in some of the first territories annexed into the new nation following its independence. Other military facilities, constructed in the twentieth century and simply built on available open land rather than existing Army forts and Navy yards, began making their history when the airplanes arrived.

The states of Texas, Louisiana, Mississippi, Alabama, Florida, and, for the purpose of this book, Georgia, provide the United States with a unique safe haven of open gulf water and unrestricted skies for training and defensive operations. Gulf-state air bases have fueled local economies from the moment of their inception, over time generating millions of dollars and providing thousands of civilian jobs across the region. They have also trained and produced some of America's most famous military aviators.

In contemporary times, military aviation's almost instant global access has become one of this nation's most critical tools for the defense of freedom. In emergencies, troops and supplies are mobilized within hours to points around the world from bases having a long and colorful history. This book explores that unique history. Each chapter presents a specific era, demonstrating how threats imposed upon our nation resulted in the practice and refinement of aviation innovation.

In its first century of existence, the United States faced adversaries intent on invading the growing nation through the Gulf of Mexico, a once sparsely defended point of entry. In both diplomatic and military strategic planning throughout the eighteenth and nineteenth centuries, the vulnerability of the vast coastal territory was recognized as a significant threat to the survival of the young nation. Military aviation throughout the region functions much like forts of the eighteenth and nineteenth centuries. The skies above are now a critical part of the defense mission.

In January 1815, the coda of the War of 1812, British forces used Gulf Coast barrier islands as a staging area for invasion in a strategic miscalculation now famously known as the Battle of New Orleans. The intent was to capture the city and other Gulf Coast settlements, including Pensacola and Mobile, through flanks around the mouth of the Mississippi, then advance northward up the river valley and eastward through what was then a back door into the United States. The British anticipated little resistance before reaching and recapturing their more populated former colonies on the East Coast. The outcome proved to be much different.

This lesson in national defense remained paramount in the minds of American military strategists for generations to come. To prepare for future threats of invasion, a series of military posts, similar to those established along the Eastern Seaboard, were constructed along the Gulf Coast.

Forty-six years after the Battle of New Orleans, in the mists of civil war, Union commanders strangled the Confederacy through blockade and invasion of Gulf Coast cities such as New Orleans, Pensacola, and Mobile, as well as smaller towns. Success of the Union advance can be attributed in no small measure to the capture of these valuable coastal fortifications.

Cities and ports along the Gulf Coast were used as bases of operation during the Spanish American War at the turn of the nineteenth century. In World War I, the incorporation of the airplane as a weapon established a new direction in military tactics. Today, still serving as a strategic line of defense against threats unheard of in the times of America's founding or first century of development, the Gulf South is far more than a back door to the United States. The region is critical to training military personnel in every service branch. In turn, the U.S. armed services are an invaluable and integral part of the Gulf Coast economy, generating billions of dollars in revenues annually throughout these states.

CHAPTER ONE

FROM THE GREAT BIRTH TO THE GREAT WAR

In the years prior to, during, and immediately following World War I, the concept of human flight was considered little more than a novelty by much of the population. At that time, aviation as an activity—much less as an engineering science critical to a progressive modern civilization—had an uncertain future, to say the least. Early in the twentieth century, during the discovery of the aircraft's potential, both the U.S. Army and the U.S. Navy selected locations in the Gulf South to establish their principal training bases. As a result, and because of a climate suitable for year-round flying, the region became the nation's earliest concentrated area for military aviation training and operations.

World War I is commonly considered the occasion of the birth, or at least the first application, of military flying. However, a number of significant events occurred before that conflict, especially in the Gulf South, that marked the beginnings of military aviation history in the United States.

THE EARLIEST FLIGHTS

In the years following the experimental flights of the Wright brothers at Kitty Hawk, North Carolina, the pair embarked on a global campaign to attract customers for their Flyer. The brothers traveled the nation on promotional tours to demonstrate the aircraft, refining their aerial salesmanship over the years. High-ranking officers from each branch of the military were present at these demonstrations.

An aggressive sales blitz overseas to promote the Wright Flyer led European countries to express interest in the invention, but the tour itself generated little money. It was apparent that European nations and kingdoms were developing plans of their own for the airplane. Motivated at least in part by the foreign attention the Flyer was receiving, as well as ongoing efforts of European enthusiasts to design a craft of their own, the U.S. Army advertised a formal public bid for design and purchase of an aircraft in late 1907. A Wright airplane was selected, and its introduction and flight trials were completed in the summer of 1908. The Army took possession of the aircraft in 1909 but, rather than treating it like the discovery of a new practical field application, considered it more of a bothersome experiment.

In coordination with the Army's 1907 initiative to acquire an aircraft, a separate unit was created within the service, aptly named the Aeronautical Division of the Signal Corps. It was tasked with exploring all aviation activity, from the use of observation balloons to the feasibility of aircraft on the battlefield. By the time the Aeronautical Division was established, two competing aircraft design teams, one led by the Wright brothers and another by Glenn Curtiss, vied for the military's business.

1910. The original Wright Brothers' airplane shed constructed on the Frank Kohn cotton plantation featured a series of advertisements and publicity for sponsors. (Photo courtesy of the Air University/Air Force Historical Research Center, Maxwell Air Force Base, AL)

One of the southernmost locations selected for demonstration flights was a cotton field near Montgomery, Alabama, borrowed for the purpose from its owner, Frank Kohn. From March through May 1910, the community was home to Wilbur Wright, who ran the operation. Although the Wright brothers' presence was brief, the hangar and Alabama's newfound interest in flying were destined to become a prominent fixture in the history of military aviation.

After a number of experimental flights utilizing available fields throughout the country, the single Army aircraft was transported by railroad to Fort Sam Houston, outside of San Antonio, Texas, where it was received with skepticism as a waste of valuable funds. That the fort was to become a base of aviation operations disgusted senior officers, who allowed exploration and testing of the airplane to be carried out with a great deal of independence and the caveat that any applications discovered by those few experiencing the thrills of flying be presented for further study. Army personnel assigned to the project resolved that the airplane would advance field observation for artillery and infantry movement.

The military heritage of this hallowed Texas ground goes back to 1845, when in the decade following a hard-fought independence from Mexico, it was an outpost and supply depot. The site became known as the Quadrangle, with its buildings constructed from limestone mined out of a nearby quarry. Major events that followed, including the annexation of Texas by the United States, statehood itself, the Civil War, and Reconstruction, all influenced the development of the city of San Antonio and the military installations surrounding it. Designated in 1890 as Fort Sam Houston, in honor of one of the most revered Texans of all time, the site continued to play a vital role in the lives of the growing western population, becoming one of the largest forts operated by the U.S. Army.

With the aircraft designated as U.S. Army Signal Corps Airplane No. 1, an assigned officer, 1st Lt. Benjamin Foulois, taught himself to fly most impressively, without the assistance of an instructor present at the fort. Modifications made to the aircraft included the attachment of wheels—a first in aviation—to supplement the skids provided in the Wright brothers' original design.

As interest grew and the potential for aviation was realized, the Army ordered additional planes. The first addition was the Curtiss Type IV Military "Pusher" aircraft, described as such because the motor and propeller

1910. Lt. Benjamin D. Foulois, the first U.S. Army pilot, prepares for flight at Fort Sam Houston, Texas. (U.S. Air Force photo)

1910. Army personnel and curious onlookers prepare to watch Lieutenant Foulois pilot the Wright aircraft above Fort Sam Houston. (U.S. Air Force photo)

were behind the pilot. The program continued to develop, and in 1913, the 1st Aero Squadron, U.S. Army Signal Corps, was officially organized.

THE U.S. ARMY SIGNAL CORPS, AERONAUTICAL SECTION

In 1914, the unit showed promising developments. The Aeronautical Division was expanded into its own "Section," providing additional independence and resources. Another aircraft model was soon added to the aviation inventory, the Jenny, a newer design provided by Glenn Curtiss. Among its features was an engine mounted in front of the pilot, tandem seating for pilot and a single passenger, and a single tail and elevator mechanism, eliminating much of the complex wiring and rigging required on previous pusher models. Following this successful introduction and advances in the program, the Army initiated steps to develop a permanent pilot and aviation infrastructure incorporating land and forts near San Antonio, utilizing the central and Gulf Coast regions of Texas as a training area.

In March 1916, a revolution in Mexico reached the border of the United States, resulting in armed exchanges between the two nations. When attacks by Mexican rebels extended into border towns in Texas and New Mexico, the Signal Corps Aeronautical Section fell upon its inaugural combat assignment. The Aeronautical Section was used to support the infantry responding to insurgency. Army pilots and aircraft flew into Mexico to observe and report the movements of Pancho Villa, the infamous leader of the revolution. These flights became the first true combat air missions ever conducted by the United States Army.

In April 1917, the same month that the United States entered the conflict engulfing Europe, the Army commissioned Kelly Field in honor of Lt. George Kelly, an Army pilot recently killed in a flying accident. The mission of Kelly Field was to consolidate Army aviation training, which to this point was scattered geographically. Additional Aero squadrons were formed as the infrastructure at Kelly expanded with demand. The primary airfield was named Kelly Field No. 1, and new outlying airfields were also named by assigned number—such as Kelly Field No. 2—as the operation developed and further specialized. Shortly thereafter, one of these outlying fields was renamed Brooks Field in honor of Cadet Sidney L. Brooks, a San Antonio native who became the first training fatality at the Kelly Field operation.

1918. Curtiss Jennys above Kelly Field. (U.S. Air Force photo)

1923. Front entrance to Kelly Field No. 1 (*left*) and Kelly Field No. 2 (*right*) at the time of World War I. (U.S. Air Force photo)

1923. Aerial view of Kelly No. 1. (U.S. Air Force photo)

THE FOUNDATIONS OF NAVAL AVIATION ON THE GULF COAST

The U.S. Navy owes a better part of its aviation history to the Gulf Coast communities surrounding the city of Pensacola, Florida, and Escambia and Pensacola Bays. Naval Air Station Pensacola, the first naval air station operated by the United States, opened in 1914 on the remnants of the historic Pensacola Navy Yard, which was commissioned by President John Quincy Adams in 1826, five years after Florida became a U.S. territory.

With the Battle of New Orleans a vibrant memory less than a decade old, and with European instability still a threat as the victors of the Napoleonic Wars remapped that continent and its colonial wealth, the president and his cabinet capitalized on every opportunity to defend this active southern entrance into the young nation. For nearly a century, the Pensacola Navy Yard constructed naval warships and repaired and provided safe harbor for vessels patrolling the volatile waters of the Caribbean and Gulf of Mexico.

Construction of Fort Barrancas soon followed, one of a number of forts built along the Gulf Coast shoreline after the War of 1812. The site was a bluff, or *barranca*, that since the 1700s had held numerous Spanish colonial fortifications, French occupational forces, and British redoubts. An arsenal supplied the weaponry to defend the new Navy yard and the surrounding Florida coastline.

During the Civil War, the Pensacola yard was occupied in turn by Confederate and Union troops. Confederates captured, then abandoned the yard following Union success in the Gulf campaigns, including the capture of New Orleans and Mobile. In the course of this exchange, many of the original yard buildings were destroyed.

In 1906, a number of replacement buildings were damaged by a hurricane and storm surge. As rebuilding continued, an outbreak of yellow fever in the area further delayed a full recovery. With the pace slow, Secretary of the Navy Josephus Daniels decommissioned Pensacola Navy Yard in October 1911.

Glenn Curtiss continued his pursuit of a military contract for aviation and believed that the U.S. Navy would find a practical use for flying. He was convinced that, like the Army, the Navy would conduct research on adopting the airplane for its own purposes.

In January 1911, San Francisco Bay, California, hosted the first demonstration of an aircraft landing on a ship, the USS *Pennsylvania*.

1910. Lt. Theodore "Spuds" Ellyson (*left*) was the first naval aviator trained by Glenn Curtiss (*right*). (Photo courtesy of the Emil Buehler Library, National Naval Aviation Museum, NAS Pensacola, FL)

Top: **1918.** The property near Opa-locka, Florida, used by Glenn Curtiss later became Naval Air Station Miami. (Photo courtesy of the Emil Buehler Library, National Naval Aviation Museum, NAS Pensacola, FL)

Above left: **1913.** Lt. Fred Seydel and Lt. W. C. Sherman (*seated in the aircraft*) of the U.S. Army Signal Corps First Aero Squadron operating Wright aircraft No. 11 in Texas City, Texas. (U.S. Air Force photo)

Above right: **1915.** This photo shows John Towers being carried to his Curtiss flying boat in preparation for launch at NAS Pensacola. Pilots were often carried to their aircraft positioned in the water for takeoff to keep their boots and pants dry for the flight. (Photo courtesy of the Emil Buehler Library, National Naval Aviation Museum, NAS Pensacola, FL)

1910. The earliest hangars constructed at NAS Pensacola were tents. (Photo courtesy of the Public Affairs Office, NAS Pensacola)

Executed on a short, makeshift flight deck, the landing was assisted by primitive arresting wires made of sandbags and rope, and was followed approximately one hour later by a takeoff. This was the second takeoff from a ship, the first conducted on the USS *Birmingham* in November 1910 on the Elizabeth River in Virginia. Both flights used a Curtiss Pusher piloted by Eugene Ely, an associate of Curtiss.

Congress passed the Navy Appropriation Act in 1912, providing for the development of naval aviation. Following the act, three aircraft—one Wright Flyer and two Curtiss Pushers—were purchased by the Navy. As a part of this new aeronautical integration, both Wright and Curtiss trained a small group of naval officers on their planes.

The naval officers, intent on incorporating aviation into the fleet, expressed interest in Curtiss's design for a floatplane, an aircraft of similar style but adapted for landing on water with "floats." Curtiss demonstrated his newest models near Miami. The Navy purchased a handful of these aircraft and proceeded to explore how to use this new technology in its operation.

In October 1913, Secretary Daniels commissioned a board to consider options for establishing a permanent naval aviation base for training and operations. The old Pensacola Navy Yard, with its vast acreage and abundant bay waters adjacent to the Gulf of Mexico, was recommended. By the beginning of the twentieth century, the area provided a secure, isolated southern Gulf base located at the tip of the Florida panhandle. As utilization of the airplane became a reality, naval officers saw the Florida Gulf Coast as a perfect inland environment for experimental seaplane operations. The recommendation was approved, and in 1914, Naval Air Station Pensacola was commissioned on the grounds of the old Pensacola Navy Yard.

In contrast to the Army Signal Corps, the Navy planned to utilize aircraft for defense of open water and to deliver pilots for fleet action with missions over the shorelines and oceans of the globe. Embarking on a mission and skills much different than those of their land-based Army counterpart in Texas, the Navy developed a unique flight training program along the Florida Gulf Coast. The Pensacola pilots, referred to as "naval aviators," were challenged to develop this newest asset for the fleet.

Station construction began with refurbished quarters and support buildings for the new personnel, a small unit of officers and mechanics. Tents were used as temporary hangars erected along the southernmost shore of the station on Escambia Bay until a permanent plan was devised. Wooden ramps were installed in front of each of the tents to ease movement of the aircraft through the beach sand. Eight additional Curtiss floatplanes were delivered to the air station.

1915. Curtiss flying boats taking off from the Florida bay. (Photo courtesy of the Emil Buehler Library, National Naval Aviation Museum, NAS Pensacola, FL)

WORLD WAR I

The cornerstone event of this era was the global conflict first known as the Great War, which began among European adversaries on July 28, 1914, and later was known as World War I, ending with the Armistice of November 11, 1918. The United States did not enter the war until 1917, long after hostilities had engulfed the European continent and its breadth of political dominion across the world.

Although World War I was the first global conflict to use the airplane as a weapon in battle, flight itself was not new to the warfront. As early as the French Revolution (1789) to 1815, observation balloons were used to rise above terrain obstacles and view enemy movements from as far as a mile away. The aircraft became a significant tool and weapon in the European theater in World War I, but in the United States, its importance was so undervalued that few airplanes were transported to Europe following the declaration of war. As a result, American pilots, including the renowned Eddie Rickenbacker, made their name in aircraft received from the Allies.

The growing number of those who believed in the potential of the aircraft had the mission of persuading a skeptical command. In the U.S. military infrastructure, many high-ranking generals and admirals were

Above: This 1917 photo shows that airships were introduced to Navy air operations prior to the U.S. entrance into World War I. This floating airship hangar was later permanently relocated onto land at Station Field, NAS Pensacola. See photo on page 30. (Photo courtesy of the Public Affairs Office, NAS Pensacola)

Below: A 1916 pre-war concept drawing of the planned layout for hangar construction at NAS Pensacola. (Photo courtesy of the Emil Buehler Library, National Naval Aviation Museum, NAS Pensacola, FL)

In sharp contrast to earlier floatplane designs, these 1917 and 1918 Curtiss H-12L models at NAS Pensacola reveal the advances being incorporated into flight operations by naval aviators. (Photo courtesy of the Emil Buehler Library, National Naval Aviation Museum, NAS Pensacola, FL)

Aircraft of the 135th Aero Squadron line up on August 7, 1918, only a month before war's end, for the first mission flown over the front by U.S.-built DH-4s. (U.S. Air Force photo)

reluctant to give advocates of the airplane a primary rather than supporting role. Others believed. Seeing the possibilities, they hinted that it was an innovative weapon from which an entirely new branch of service could be established.

In contrast to this new flight activity, Gulf Coast cities such as New Orleans, Louisiana, Gulfport, Mississippi, and Mobile, Alabama, were major ports of embarkment. New Orleans was so laden with military troops preparing for transport to Europe that Storyville, the infamous red light district behind the French Quarter, was closed at the request of senior military officers concerned with the well-being of the troops.

These central Gulf Coastal communities remained traditional military hosts, with only an occasional aircraft landing in a nearby open field. An exception was the Navy seaplanes flying from the new naval air station in Pensacola to the naval base in Algiers, Louisiana, located directly across the Mississippi River from New Orleans. Pilots made takeoffs and landings on the river, mooring along the banks. Because of its rarity, American military flying outside of airfield areas prior to and during World War I was perceived by a better portion of the general public as a grand experiment.

After the war, the need for pilot training slowed considerably in both the Army and the Navy. Kelly curtailed its growth, and Brooks Field stopped training operations. For years to come, NAS Pensacola remained the only base of its kind, exclusively designated as an aviation operation for the Navy. However, with the multitude of pilots returning from the European theater, it was apparent that there was a need for expanded bases of operations that could utilize the experience of these seasoned flyers.

In 1918, the Navy incorporated the large twin-engine flying boat the Curtiss F-5L into its fleet. The aircraft could accommodate up to ten crew members and remained in service for ten years. In May 1919, a Curtiss aircraft of similar design, the NC-4, completed the first trans-Atlantic crossing by air (though not a non-stop), eight years prior to the famous flight by Charles Lindbergh. (Photo courtesy of the Emil Buehler Library, National Naval Aviation Museum, NAS Pensacola, FL)

Montgomery Air Depot in 1919, after World War I ended. The base was later renamed Maxwell Field. (Photo courtesy of the Air University/Air Force Historical Research Center, Maxwell Air Force Base, AL)

NAS Pensacola, 1918. (Photo courtesy of the Emil Buehler Library, National Naval Aviation Museum, NAS Pensacola, FL)

THE PENSACOLA NAVY YARD AND THE NAVAL AIR STATION HISTORIC DISTRICT

Following his inauguration as president of the United States in 1825, one of the first actions of John Quincy Adams, son of founding father John Adams, was to authorize the building of a naval yard in Pensacola, Florida. Acquired from Spain in 1819 while Adams was secretary of state, Florida was a volatile and unstable colonial region. Colonial Florida was divided into two parts: West Florida, which included the panhandle where the strategic settlements of Biloxi, Mobile, and Pensacola were located; and East Florida, which spanned the remaining peninsular lands stretching to the Atlantic on the east and the southernmost Gulf Coast on the west. The area was in dire need of a permanent force capable of suppressing threats ranging from the pirate black market and slave trade operations to conflicts with Native Americans and the potential for resurgence of European strength along the Gulf Coast.

The Pensacola Navy Yard became the first United States Naval post in the Gulf Coast region. With abundant forests and timber nearby, much-needed naval vessels were constructed in the Pensacola yard; ironically, some were destined to fight on opposing sides in the Civil War just a few years away. Despite all of this activity driven by Adams and his successors, Florida did not become a state until 1845.

Over the course of the next century and a half, multiple buildings and fortifications (estimates as high as three hundred in number) were constructed on the historic property of what is now Naval Air Station Pensacola. The ravages of time and natural disasters—including as many as ten hurricanes making near direct hits on Pensacola Bay and Escambia County—damaged or completely destroyed these historic properties. In December 1976, the Pensacola Naval Air Station

Opposite: An early artist's rendering of the Pensacola Navy Yard following commission, the first U.S. Navy property later dedicated to aviation. (Photo courtesy of the Public Affairs Office, NAS Pensacola)

Above: The NAS Pensacola officers' quarters in 1918 from the view of a Navy airship. (Photo courtesy of the Public Affairs Office, NAS Pensacola)

Historic District was established by the U.S. National Park Service to preserve and maintain the buildings that remained.

Irreversible devastation occurred in September 2004, when Hurricane Ivan destroyed approximately 90 percent of the historic district, a significant loss not only for the U.S. Navy, but for our national heritage. Of the buildings damaged or destroyed, 37 were on the National Register of Historic Places. Gone are the first aircraft hangars built to replace the original canvas tents used to house aircraft; gone are the original quartermaster offices and various other structures along the shore of Pensacola Bay used by the first naval aviator classes that graduated from the air station.

Fortunately, some historic properties associated with NAS Pensacola's pioneering days remain. Still standing are the original Admiral's House and a number of the officers' homes, including that of Admiral William "Bull" Halsey when he was in command of the air station. The majority of these buildings are located farther inland rather than immediately adjacent to the bay. Although tours of the base are available through the air station's Public Affairs Office, a number of these surviving historic buildings remain private residences for the base's officers or serve as facilities not currently open to the public.

1930. Tactical formation of nine attack aircraft from Kelly Field, Texas. (U.S. Air Force photo)

CHAPTER TWO

BETWEEN THE WORLD WARS

The twenty-three years of peace experienced by the United States between 1918 and 1941 influenced the development of military aviation as much as the world wars themselves. Unfortunately, nearly all of the innovations of this productive yet underappreciated period have been overshadowed by the decades that preceded and followed.

A range of experiences—from the welcome return of the American Expeditionary Forces (AEF) to alarm at Adolf Hitler's transformation of Germany—in addition to a national desire to be free from foreign entanglements fueled amazing innovation in American aircraft design. Within this period, however, the Great Depression and President Franklin D. Roosevelt's recovery programs preoccupied the American mind, leaving little time for the average citizen to think about flying or other matters. Precious tax dollars necessary to capitalize on such inspirations were fought over by proponents of emergency economic recovery programs, often at the expense of military aviation and national defense. The result can best be described as a slow period of transition by a nation at peace, struggling to maintain its isolation within a world moving once again toward global conflict.

In the face of these challenges, military flying continued to develop, especially in the Gulf South states. New facilities introduced a creative blend of local architecture, a unique home for the rapidly advancing technology that in a few short years would help the United States face a global crisis.

FOUNDING OF THE U.S. ARMY AIR CORPS

Aviation's evolution into an efficient, operational Army asset resulted from numerous primitive flight experiments across the nation. Regular military command reorganizations accompanied this development. As a result, laypersons studying the history of military aviation find themselves greeted with a complex family tree of early military flying branches. Frequent redefinition was common as the benefits and promise of aviation were discovered and the most practical means of application were implemented.

The Army Air Service units that so ably brought flight into practical use and supported the AEF in World War I had grown into viable training and operational units. By the 1920s, their expanding use required yet another distinction. In 1926, Congress passed the U.S. Army Air Corps Act, repositioning the Army Air Service into a separate command named the Army Air Corps. This new stature stopped short of complete independence from the Army, but it provided military aviation with the growing room necessary to build a solid foundation and to compete on the global stage.

From the early 1930s through the end of World War II, both Army Air Corps and naval aviation cadets received primary pilot training in the Stearman (Boeing) N2S at bases throughout the Gulf South. (U.S. Air Force photo)

As a part of this restructuring, the missions of military aviation were narrowed and defined. Flight training was consolidated. Specialized squadrons emphasizing bombing, pursuit (later reclassed as fighter), and observation were further expanded to concentrate on these flying skills.

The Air Corps Act established the Air Corps Training Center at Kelly Field No. 1, renamed Duncan Field shortly thereafter. The School of Aviation Medicine, which had been established at Hazelhurst Field, New York, in 1918, was relocated to another of the original Kelly stations, Brooks Field, in 1926. Although this consolidation would provide much-needed training and direction for the Air Corps, the facilities the Kelly fields offered were fast becoming inadequate. In addition to the problem of limited acreage, the buildings were deteriorating, being in most cases the remnants of facilities hastily constructed in response to World War I.

Besides the San Antonio area, the Army continued to operate airfields in other parts of Texas, including Ellington Field in Houston, Taliaferro Field near Fort Worth, Love Field in Dallas, Barron Field in Everman, Hicks Field in Saginaw, Carruthers Field near Benbrook, Call Field in Wichita Falls, Rich Field in Waco, and Fort Crockett near Galveston. These smaller facilities were home to air squadrons and supported the Kelly operations as remote and outlying airfields for pilots to conduct training. While other states had a number of airfields, it was this particular concentration in Texas that led to a critical decision that would change the future of Army flight training in the Gulf South.

RANDOLPH FIELD, TEXAS

As the limitations of the original Kelly airfields and infrastructure steadily revealed themselves to the military command, a new direction for Army aviation was set in motion. General Frank Lahm, known and respected as the first official military pilot, had been assigned to lead the new Air Corps training program. General Lahm's vision was to continue training in the San Antonio area, integrating the old bases into plans for an entirely new, masterfully designed air facility.

Site selection quickly focused on open acreage approximately fifteen miles from San Antonio. The city acquired the land for the sole purpose of donating it to the U.S. government for use as a military base. Texas congressman John Nance Garner, a member of the U.S. House of Representatives from 1903 to 1933, facilitated selection and development of the site. The congressman's powerful presence in Washington, D.C., prompted President Roosevelt to select him to serve as vice-president from 1933 to 1941.

1931. The original water tower that forms the centerpiece of historic Randolph Air Force Base. (U.S. Air Force photo, Tarrant County College NE, Heritage Room, Portal to Texas History)

A team chaired by General Lahm considered a number of designs for the new facility. The favored concept came from an unexpected though well-informed source, Harold L. Clark, a young lieutenant assigned to the motor pool at Kelly Field. Lieutenant Clark had a unique pastime. Prior to enlisting in the Army, he studied architecture in college. Witnessing aviation develop around him, he began sketching layouts of airfields as he imagined them, including spaces meeting the needs of every associated support, from hangars and runways to housing for personnel and administrative offices.

Clark designed a circular pattern of roadways that formed a self-sustaining city with landing strips on three sides and a single main entrance avenue for access to the facility. Within the circle was a vast area of real estate with plentiful space for future growth. The design allowed for strategic placement of specialized infrastructure, such as aircraft storage and hangars, mechanical support, and training facilities. Clark offered his drawings to General Lahm, who immediately saw the potential of the idea. Presented to a board previously commissioned to select a layout for the base, the lieutenant's concept was approved. After further development, Clark's design was delivered to local firms for refinement and implementation.

Next, an architectural style was chosen for the new facility. The Army was in the early stages of introducing two styles for new facilities being constructed on bases throughout the country: a Colonial Revival style for bases in the northern states, and a Spanish Colonial Revival style for bases in the southern and western states. Thus, the new San Antonio airfield would reflect the Spanish Colonial style.

The layout of the entrance avenue was designed to offer a grand first impression to personnel and visitors upon their arrival at this vast new home for Army aviation. The avenue features some of the facility's most prominent buildings and leads to Washington Circle, which honors the 200th anniversary of the first president's birth. As the central axis of the base to this day, the avenue connects a network of thirty-one miles of roadways throughout the site.

Positioned at the end of the entrance avenue is the architectural centerpiece of the base, a towering landmark destined to become known to generations of personnel as the "Taj Mahal." The origins of the reference are informal, the building's style having little resemblance to the masterpiece of India with the exception of its dome and intimidating height above surrounding structures. It is home to administrative offices and encases the original base water tower, crowned with tile in the Spanish mission style. Among the buildings near the "Taj" is the base chapel, which takes the heritage of the Spanish Colonial style to even greater symbolic heart. The twin bell towers do not match, in keeping with the tradition of never completing a chapel in order to avoid taxation.

1931. Observing a Spanish Colonial tradition, a single steeple on the chapel at Randolph remains unfinished. (Photo courtesy of JBSA Public Affairs Office)

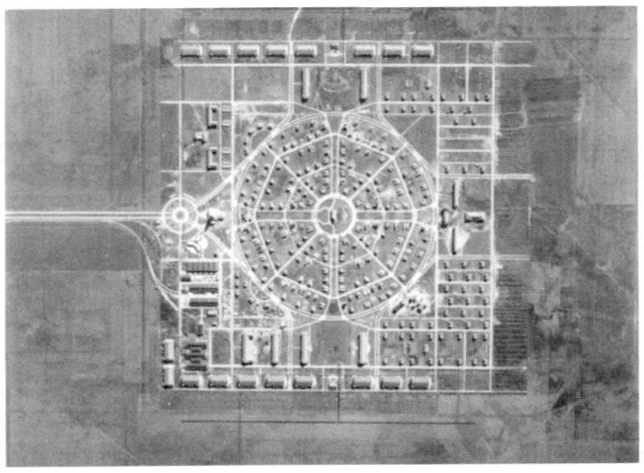

Left: **1931.** This aerial view of Randolph Field reveals the "circle within the square" design conceived by Lt. Harold Clark. The entrance road and water tower are seen at the left center of the photo. (U.S. Air Force photo, Tarrant County College NE, Heritage Room, Portal to Texas History)

Right: **1931.** Construction of Washington Circle near the water tower. (U.S. Air Force photo, Tarrant County College NE, Heritage Room, Portal to Texas History)

1935. The chapel (*left*), administrative buildings (*background*), and tower (*right*) at Randolph Field. (U.S. Air Force photo, Tarrant County College NE, Heritage Room, Portal to Texas History)

1934. The barracks, designed in classic Spanish Colonial style. (U.S. Air Force photo, Tarrant County College NE, Heritage Room, Portal to Texas History)

1935. The base hospital at Randolph. (U.S. Air Force photo, Tarrant County College NE, Heritage Room, Portal to Texas History)

Much like Kelly, the name for the base made a tragically abrupt appearance. Captain William Randolph, a member of the committee assigned to name the new facility, was killed in a plane crash on February 17, 1928. Within six months, the committee selected the name Randolph Field in his honor.

Randolph Field's original land base extended over 2,300 acres, all dedicated to the aviation project. From the outset, even as the operation grew, there was minimal danger of overcrowding the property with a dense population. The series of original buildings on the base, identified by number and later immortalized by the names of distinguished service personnel, were constructed throughout the circled streets in near perfect geographic placement.

Rows of hangars—featuring up to ten structures per side on the east and west parameters of the circle—were built to support the airfields. In addition to the high volume of training aircraft moving onto the field and related maintenance support required for these growing fleets, these hangar rows housed Link Trainers, the first flight simulators created to introduce student pilots to instrument flying.

Randolph Field was officially dedicated on June 20, 1930. It employed thousands of military and civilian personnel, becoming one of the most prominent military properties in the United States and carrying on the tradition of the original Kelly bases that surrounded it. By 1931, the operation had become so large that it was nicknamed the "West Point of the Air," a fitting tribute for the base that would train or influence nearly every U.S. Army pilot for decades to come.

BARKSDALE FIELD, LOUISIANA

As Randolph Field developed, the Air Corps continued to expand military aviation's footprint, searching for other areas in the country to build airfields. In September 1927, the War Department began looking

Top left: **1936.** An aerial photograph of a fighter group line inspection on the ramp at Barksdale. (Photo courtesy of the Barksdale Global Power Museum, 2nd Bomb Wing)

Top right: **1934.** Barksdale Army Air Field, Louisiana, soon after its opening. (Photo courtesy of the Barksdale Global Power Museum, 2nd Bomb Wing)

Bottom: **1937.** The Eighth Aero Squadron at Barksdale Field. (Photo courtesy of the Barksdale Global Power Museum, 2nd Bomb Wing)

This 1943 photograph highlights the French Colonial elements incorporated into the design of Barksdale Field to reflect Louisiana's colonial heritage. (Photo courtesy of the Barksdale Global Power Museum, 2nd Bomb Wing)

1934. Aerial view of recently completed Barksdale Field. (Photo courtesy of the Barksdale Global Power Museum, 2nd Bomb Wing)

1934. "Open House" public air show on Barksdale Field. (Photo courtesy of the Barksdale Global Power Museum, 2nd Bomb Wing)

for additional acreage to establish active air bases that could be staffed with pilots and flight crews trained at Randolph to carry out specific missions. Among the many places competing to host this base was the metropolitan area of Bossier City–Shreveport, Louisiana.

After winning a national competition, these northwest Louisiana municipalities issued bonds to purchase land selected through a flyover conducted by local pilot Harold R. Harris. Construction of the new base began in March 1931, with buildings designed in a modified Spanish Revival style that incorporated French Colonial elements, thus reflecting the Louisiana culture. The base was named after Lt. Eugene H. Barksdale, a veteran pilot of the World War I American Expeditionary Forces who was killed in a crash on August 11, 1926. Unlike Kelly and Randolph, Lieutenant Barksdale died prior to identification of the land that would soon become home to his legacy.

Barksdale Army Air Field was dedicated on February 2, 1933, just under two years from the date construction began. The base design, though less intricate than the Randolph circle layout, featured all of the support infrastructure necessary to fulfill the mission. Like Randolph's impact on San Antonio, the sudden influx of military activity at Barksdale provided a welcome boost to local and state economies during one of the most pervasive economic depressions the nation had experienced.

Barksdale was part of the developing larger plan for a network of new Air Corps training and operational facilities. In the first phase following the 1926 Air Corps Act, Army General Headquarters (GHQ) established three combat wings. In March 1935, the Third General Headquarters was commissioned at Barksdale's new facility. Supporting the center of the nation, Barksdale was accompanied by two similar Air Corps bases located along the nation's coastlines. Langley Field in Virginia was expanded to support operational squadrons in the eastern United States. March Field, located near Riverside, California, became the headquarters for Air Corps operations on the West Coast.

As a result, a number of active pursuit and bomber squadrons were relocated to Barksdale Field. In addition, new squadrons were established to develop tactical and support techniques for the rapidly evolving discipline of air warfare across the globe. One such unit was the Southeast Air Corps Training Center, created for specialized advanced training of multi-engine crews, including pilots and navigators, following primary and advanced training at Randolph.

Top: Maxwell Field in the late 1920s. (Photo courtesy of the Air University/Air Force Historical Research Center, Maxwell Air Force Base, AL)

Bottom: Some of the earliest military photo training was conducted at Maxwell Field in the 1920s. The technical craft of aerial photography was advanced in the years between the world wars. (Photo courtesy of the Air University/Air Force Historical Research Center, Maxwell Air Force Base, AL)

Opposite: This rare 1920s aerial photo shows Pensacola Bay's multiple seaplane training operations underway. (Photo courtesy of the Public Affairs Office, NAS Pensacola)

MAXWELL FIELD, ALABAMA

Having served as one of the southernmost locations selected for promoting aircraft in its infancy in America, Frank Kohn's Alabama cotton field was leased by the Army years after the Wright brothers' departure and used as a repair facility for aircraft throughout World War I. The Army later purchased the property. Although threatened by the harsh realities of the postwar economic downturn, the operation held on through 1920, when its future namesake, 2nd Lt. William Maxwell, a native of Atmore, Alabama, lost his life in a flying accident in the Philippines. Two years later, the Alabama base was renamed Maxwell Field in his honor.

Expansion of Maxwell in the late 1920s coincided with the ongoing development of aviation across the nation. The Army Air Corps was establishing a broad infrastructure to maintain specific, rapidly evolving flight disciplines, including tactical operations, pursuit, and observation. A number of states, especially those along the Gulf Coast, were competing to attract squadrons and military training schools to their municipalities. Maxwell was selected as the new home of the Army Air Corps Tactical School (ACTS), an academic unit founded in 1920 at Langley Field, Virginia. With its specialty in tactical operations, the base became a center for developing aerial bombing and the support required to maintain a growing fleet of larger aircraft suitable for the mission.

Maxwell Field—like Randolph, Barksdale, and a host of other bases with increasingly specialized aviation missions and disciplines—received funding to replace obsolete buildings remaining from World War I and to construct entirely new facilities to house the influx of personnel and equipment. Among the first of these new structures was Austin Hall, which opened in 1931 as home of the tactical school and was named in honor of 1st Lt. Charles Austin, an instructor at the school's original location who died from an illness in 1928.

Maxwell Field's growing mission prompted the development of other airfields in surrounding states. One of the more prominent was the Valparaiso Bombing and Gunnery Range, located to the south, outside the Florida city of the same name. Training on Maxwell Field was expanded in 1935 to incorporate this location, favored because of its isolation and vast forest acreage beside the Gulf of Mexico that was suitable for live bombing practice.

NAVAL AIR STATION PENSACOLA, FLORIDA: A RENEWED MISSION

To this day and with good reason, NAS Pensacola is fondly referred to as the "Annapolis of the Air" and the "Cradle of Naval Aviation." Naval aviation owes much of its principal development to this part of the Gulf Coast. The Pensacola area quickly became a magnet, attracting flight activity from other parts of the nation to the Florida panhandle.

Pensacola's advantages for naval aviation are many. The region contains inland bays with plentiful opportunities for training. While the area did host considerable shipping activity in the early twentieth century, it was at the time relatively isolated, if not remote. Most commercial shipping migrated to the heavier sea lanes leading to the Mississippi River and ports along the East Coast. In addition, the infrastructure of roadways through the area's heavily forested shores remained in place through the early 1920s and afforded a foundation for continued development. As a result, Pensacola and the Escambia Bay area offered the naval aviation program unique, undisturbed open-water access and room to grow.

NAS Pensacola became the center of spirited debate: What would become the most practical means for naval air operations: aircraft operating from land-based fields, or "flying boats"? One matter was certain: regardless of the vehicle carrying them aloft, aviators provided the Navy's fleet with the ultimate crow's-nest view.

The Pensacola station began operations without a means to support land-based aircraft. Instead, from its inception, the station employed methods pioneered in the United States by Glenn Curtiss, whereby fuselage hulls and pontoons replaced early skids and wheels used by civilian and Army pilots. The ability to take off and land in water in support of the Navy fleet was thought by Curtiss's flying boat advocates to be the primary value of naval aviation. The Coast Guard also incorporated flying boats in its patrol operations.

Another aeronautical addition to the Navy fleet was the dirigible, or lighter-than-air blimp, also known as an airship. The cumbersome aircraft lacked speed but showed promise as an escort for over-water missions. Evolving from stationary observation balloons used for the past two centuries, dirigibles were extensively refined for piloting in

1922. The wooded and swampy inland area northeast of the original Pensacola Navy Yard, including the former bay town of Woosley, was cleared and filled to build Station Field. The original location of the 90 degree–angled Navy yard brick fence is clearly seen alongside buildings that survived into the twentieth century. Just above the northeast edge of the fence (*top right*), the World War I–era oval-shaped floating airship mobile hangar can be seen, permanently relocated on the Station Field site, accompanied by a larger new hangar constructed to its left. Navy seaplane operations continued on the banks of Pensacola Bay (*lower left*). (Photo courtesy of the Public Affairs Office, NAS Pensacola)

Ca. 1930. Aircraft parked at the hangar on Station Field. In 1935, Station Field was renamed Chevalier Field in honor of Lt. Comm. Godfrey Chevalier, the first naval aviator to land on the USS *Langley.* Chevalier was killed in a training accident in November 1922, less than a month after his historic landing. (Photo courtesy of the Public Affairs Office, NAS Pensacola)

Europe, most famously by the German Baron von Zeppelin. Early Navy dirigibles remained above the fleet and reported positions of both submerged and surfaced enemy vessels. Although airship activity was more prominent along the East and West Coasts of the United States, the new Pensacola base supported an active operation over the Gulf.

Four years after the Great War, NAS Pensacola embarked on a third mission. The USS *Langley* and future carriers would depend on the Florida station to provide trained aviators for its aircraft. To the northeast of the base's operation for flying boats, beside the hangars constructed for airship operations, an airfield was constructed for a growing inventory of land-based naval aircraft that would supply carrier-trained aviators. This first land-based airfield, named Station Field, was constructed over an undeveloped Navy yard and lake where in the nineteenth century fresh timber was floated and stored for construction of wooden ships.

Presented with the capacities of the flying boat, the airship, and land-based aircraft, naval personnel explored which would be the most effective for the Navy. The answer was a combination of the three. Few blimps at the time afforded the dexterity to conduct long-range patrols over the oceans. Floatplanes had even less range or operational capability. Therefore, land-based aircraft operating from carriers at sea now demanded a place in the hierarchy.

A surge in construction of new buildings on the base featured a mix of Spanish and American Colonial-style architecture complementing the Florida landscape and city of Pensacola. However, within a short time, as the 1930s drew to a close, dramatic changes in fortune of overseas allies would overwhelm the ability of NAS Pensacola to prepare trained pilots for the fleet. The need for support from other locations would become evident.

1923. The USS *Langley* docked at NAS Pensacola beside Station Field. The first operational U.S. aircraft carrier, the *Langley* was recommissioned in 1922 from a coal carrier (USS *Jupiter*). (Photo courtesy of the Emil Buehler Library, National Naval Aviation Museum, NAS Pensacola, FL)

A Curtiss floatplane flying over the Gulf of Mexico in the late 1920s. (Photo courtesy of the Emil Buehler Library, National Naval Aviation Museum, NAS Pensacola, FL)

1924. An unidentified class of floatplane aviators at NAS Pensacola. (Photo courtesy of the Emil Buehler Library, National Naval Aviation Museum, NAS Pensacola, FL)

1935. One of NAS Pensacola's graduating classes of naval and Marine aviators at Station Field. (Photo courtesy of the Emil Buehler Library, National Naval Aviation Museum, NAS Pensacola, FL)

1938. President Franklin D. Roosevelt visits NAS Pensacola. As a former assistant secretary of the Navy, FDR was an exceptional presidential guest at naval air stations. (Photo courtesy of the Public Affairs Office, NAS Pensacola)

On November 8, 1922, the War Department renamed the Montgomery Air Intermediate Depot as Maxwell Field in honor of 2nd Lt. William C. Maxwell. A native of Natchez, Alabama, Lieutenant Maxwell died on August 12, 1920, in the Philippines when his DH-4 aircraft struck a flagpole after swerving to avoid a group of children. (Photo courtesy of the Air University/Air Force Historical Research Center, Maxwell Air Force Base, AL)

Opposite: The original naval aviator class that graduated from NAS Pensacola in 1915 included, among other officers, William Corry (*standing fourth from right*), Richard Saufley (*seated first to the left*), and Kenneth Whiting (*seated third from left*). This photograph was taken in one of the shoreline hangar tents. (Photo courtesy of the Emil Buehler Library, National Naval Aviation Museum, NAS Pensacola, FL)

NAMING AIRFIELDS

The novelty of aviation played no small part in reinforcing respect for the bold flyers daring to master this weapon of war. While hundreds of pilots were killed in combat fighting above battlefields, many other gifted pilots, pioneers of their day, were victims of fatal training or operational accidents before ever engaging in the conflict. Others perished while flying in the postwar years after their return home. The memory of these fallen combatants remained with their fellow flyers, who searched for ways to recognize their ultimate sacrifice for the country. A number of these surviving peers, as well as nonflying advocates of aviation, proposed honoring the brave personnel who died by having the names of U.S. airfields changed to memorialize them.

A comprehensive roster of all such honored personnel would require a book dedicated to the subject. However, a sampling of airfields throughout the Gulf Coast states representing each branch of the military provides insight into this revered practice. Kelly, Maxwell, Barksdale, and Randolph Fields were some of the first dedicated Army air bases constructed in the Gulf Coast states. Among examples from the 1930s and early 1940s that continue the Army and Air Force tradition are Tyndall Air Force Base in Panama City, Florida, named for Lt. Benjamin Tyndall, killed in 1930; Keesler Air Force Base in Biloxi, Mississippi, named for 2nd Lt. Samuel Keesler, a native of Greenwood, Mississippi, and member of the 24th Aero Squadron who died in a German field hospital after being shot down near Verdun, France, on October 9, 1918, only one month prior to World War I's end; and Moody Air Force Base in Valdosta, Georgia, named after Maj. George Moody, who lost his life in an aircraft flight test accident. MacDill Field, near Tampa, Florida, was named for Col. Leslie MacDill, who also perished in an air accident in 1938. Other Air Force bases constructed or improved in the post–World War II era were renamed for generals and other senior officers. One such example is Chennault Air Force Base in Lake Charles, Louisiana, named for Brig. Gen. Claire Lee Chennault, who led the Flying Tigers in China.

While naval air stations are most commonly named for the cities closest to their geographic location, the airfields constructed on these bases are named in honor of Naval aviators who set a standard that contemporary pilots continue to follow. In the years leading to World War II, the Navy opened a series of outlying fields to support the training operations at NAS Pensacola. These airfields were named

in honor of some of the first aviators to graduate from Pensacola, as follows:

- Saufley Field opened north of Pensacola in 1940 as a naval auxiliary air station and was named for LTJG Richard Saufley, Naval Aviator #14. Saufley died when his Curtiss seaplane crashed on Santa Rosa Island on June 9, 1916, while attempting to set altitude and endurance records for naval aviators.
- Corry Field, also opened as a naval auxiliary air station in a suburb of Pensacola, is named for Lt. Comm. William Corry, Naval Aviator #23, who served in World War I. He died on October 6, 1920, as a result of injuries received while rescuing another aviator from an aircraft they were piloting that crashed near Hartford, Connecticut.
- Whiting Field, a naval air station northeast of Pensacola, is named for Commander Kenneth Whiting, Naval Aviator #16. Whiting was instrumental in developing and commissioning the USS *Langley* and commanded the USS *Saratoga* in the 1930s. He died of a heart attack in April 1943.
- In 1935, Station Field was renamed Chevalier Field for Lt. Comm. Godfrey Chevalier, the first aviator to land on the *Langley*. In 1954, Chevalier Field was replaced by Sherman Field to accommodate the transition to jet aircraft. Honoring Adm. Forrest Sherman, who died in 1951, the field supports current air operations at NAS Pensacola.

In the decades following World War II, Navy airfields continued to be named for naval personnel. Among them is McCain Field at NAS Meridian, Mississippi, honoring World War II admiral John McCain, a native of Teoc, Mississippi, and grandfather of distinguished naval aviator and U.S. senator John McCain of Arizona.

THREE MEN ON A FLYING TRAPEZE

Claire Lee Chennault became one of the most famous American pilots of all time through his leadership during World War II of the American Volunteer Group (AVG), the Flying Tigers, delivering powerful offensive and physiological obstacles against unchecked advances of the Japanese ravaging freely throughout China.

Prior to World War II, Chennault had a distinguished, often controversial career. He earned his wings at Kelly Field too late to see combat in World War I and was assigned a variety of duties, including flight instructor at Maxwell Field, where he concentrated on developing pursuit (later referred to as fighter) pilot tactics and training programs. He became an advocate for using pursuit aircraft to attack bomber flights, a tactic viewed with skepticism by those advancing the growing American bomber fleet. His conviction ran contrary to the belief that the greater threat to the larger aircraft was anti-aircraft artillery fire.

Chennault was one of the few instructors who appreciated that training for pursuit pilots in the 1930s had become outdated. At best, many tactics were left over from World War I. He was not shy when expressing his opinions and pushed his aircraft beyond its limits to develop air combat maneuvers.

In 1932, to promote the agility of pursuit aircraft, Chennault created one of the first flight demonstration squads, naming it "Three Men on a Flying Trapeze," and stressing through performances the need to enhance current flight instruction. Chennault recruited two fellow flight instructors, Lt. Haywood Hansell and Sgt. William Mc-

The Trapeze pilots were (*from left*) Sgt. William "Mac" McDonald, Capt. Claire Chennault, and Sgt. John Williamson. (Photo courtesy of the Air University/Air Force Historical Research Center, Maxwell Air Force Base, AL)

Opposite: "Three Men on a Flying Trapeze," one of the first military air demonstration units, formed by Capt. Claire Chennault at Maxwell Field. In 1934 and 1935, the team flew Boeing B-12 aircraft. (Photo courtesy of the Air University/Air Force Historical Research Center, Maxwell Air Force Base, AL)

Donald, based at Maxwell Field, who shared his believe that the skills of American pilots could be expanded. Using the Boeing P-12C, the Trapeze flight demonstrations included precision loops, figure eights, rolls, and a series of other techniques adapted by Chennault for the performances. The team set out to show that a pursuit aircraft could, in the hands of trained pilots, outfly and overcome any defenses of bomber squadrons. Chennault's passion did not sit well with a number of high-ranking officers, advocates of bomber fleets who believed that these larger aircraft would evolve into a state of aerial invulnerability.

The Flying Trapeze team flew in air shows throughout the country, including the National Air Races in 1934 and 1935, where Chennault and his partners gained considerable notoriety among civilian pilots. The team was disbanded in 1936 after four years of performing. Chennault retired from the Army the following year, and though it is strongly suspected that he was forced to do so because of ongoing conflicts with superior officers, he stated that his decision was for "health" reasons.

Chennault was recruited by Chinese Generalissimo Chiang Kai-shek to organize and lead his struggling air force in its battle against Japan. The association eventually resulted in the founding of the Flying Tigers. Many of Chennault's innovations were used by the Flying Tigers and significantly delayed Japanese advances on Chinese supply lines on the Burma Road in 1941 and 1942. On March 10, 1943, President Franklin D. Roosevelt appointed Chennault commander of the newly activated Fourteenth Army Air Forces of the China-India-Burma theater.

1941. Construction continues on flying boat ramps to support PBY training at NAS Corpus Christi. (Photo courtesy of the Emil Buehler Library, National Naval Aviation Museum, NAS Pensacola, FL)

CHAPTER THREE

WORLD WAR II

In the late 1930s, President Franklin Delano Roosevelt continued his New Deal and recovery programs but could no longer put aside problems overseas while attending to domestic matters. He watched Hitler's rapidly rising Nazi party overtake the European political environment. In the Far East, Japan's increasing aggression in the Pacific continually threatened his commitment to isolationism; with that country's conquest of natural resources in the region, Allied interests were on the defensive.

The United States was one of the few industrial nations possessing a home front off the battlefields. President Roosevelt famously referred to the nation as the Arsenal of Democracy. Ironically, while manufacturing this arsenal, the United States reinforced its own military with less personnel than almost every other major nation drawn into the war. A service draft already in place, though slow to increase the ranks, was supplemented by a call for volunteers to every branch of the service after the attack on Pearl Harbor in 1941, revealing yet another problem—a dramatic shortage of bases across the nation to house and train the onslaught of new recruits.

The Navy and War Departments quickly developed plans to expand old bases and construct new ones throughout the country, a proportional share of which were established in the Gulf South. From a practical strategic point of view, the reach of aviation no longer required operating near the same bays, rivers, and forts occupied by prior forces. The demand for rapid expansion led to a gradual loss of local design influence. Military facilities throughout the region (and the country) became more uniform.

Beginning in 1941 and reaching a peak by the end of 1943, new base construction continued at a rapid pace. These new facilities would supplement and extend the presence of the original bases. Each developed a mix of missions, albeit often independent and uncoordinated from one another, together building a military large enough to operate throughout the two-front global conflict against the Axis powers.

FDR'S NAVAL EXPANSION IN THE GULF SOUTH

Having served as assistant secretary of the Navy, President Roosevelt understood military needs. By 1938, sites in the Gulf South for new naval bases and air stations had been identified, and design and construction began shortly thereafter. These would supplement the Pensacola operation, which would also be expanded. Construction took place on opposite sides of the Gulf region—in Jacksonville, Florida,

and Corpus Christi, Texas, both of which were dedicated primarily to aviation.

These air bases would support basic and advanced training with additional emphasis on the critical need for ship escort and open-water patrol. Each facility would expand the utility of the Navy's new, most widely used flying boat, the PBY Catalina. In addition to its obvious use for over-water operations, one of the primary reasons that the flying boat was so heavily emphasized in the late 1930s and at the start of the war was that a large percentage of the globe was still without paved or surfaced runways. As a result, the Navy continued to develop operations with the flying boat while simultaneously incorporating new land-based and carrier aircraft designs.

The property that became NAS Jacksonville, Florida, has a rich history. Originally, at the turn of the century, it was the site of a Florida National Guard station. During World War I, it was operated by the U.S. Army as a training camp and rifle range, said to be the largest of the period. Throughout the 1920s and 1930s, it was clear that the location of the base, beside the St. Johns River and the Atlantic, would be an ideal site for an expanded military facility, specifically for naval operations. Naval Air Station Jacksonville was commissioned in October 1940, the first of three facilities in the immediate area that would also include Naval Air Station Cecil Field for land-based aircraft and Naval Station Mayport for vessels.

Located on Florida's Atlantic coast rather than in gulf waters, this

1943. The PBY flight line at NAS Jacksonville. (Photo courtesy of the Emil Buehler Library, National Naval Aviation Museum, NAS Pensacola, FL)

Opposite: **1942.** PBY-5 Catalina aircraft provide the backdrop to graduating naval officers at NAS Corpus Christi. (Photo courtesy of the Emil Buehler Library, National Naval Aviation Museum, NAS Pensacola, FL)

base provided access for aircraft and naval vessels to the East Coast and to operations tied to the European theater. Like its counterpart in Pensacola, the station at Jacksonville featured numerous ramps for launching of the larger PBYs. Surface runways for the operation of land-based aircraft were also constructed at the station for the purpose of advancing basic flight training for naval aviators. When it began in the teens and early twenties, that basic training was conducted primarily in Curtiss floatplanes at NAS Pensacola. By this time, Navy primary and advanced flight training had evolved from flying boats to incorporate ground-based training.

On a nearly identical schedule, NAS Corpus Christi was built for the same mission as its counterparts in Florida. Once again, the powerful force of Texas politics, specifically the influence of now Vice-President John Nance Garner, played a significant role in the location and development of a dedicated naval air station on the Texas Gulf Coast. The site was selected in 1938 and commissioned for operation in March 1941. It was home to primary, basic, and advanced aviation programs.

The mission of these two new naval air stations was complementary. However, one of the principal tasks of NAS Corpus Christi was the basic training of PBY flying boat pilots in the waters of the Gulf of Mexico, while NAS Jacksonville handled more advanced instruction in the aircraft. Another station was later constructed in the vicinity of Miami, Florida, in the city of Opa-locka, where Glenn Curtiss operated a shoreline flight-training site for his flying boats. This facility was commissioned as NAS Miami in late 1940. It supported advanced training operations for naval aviators in disciplines ranging from dive and patrol bombing to fighter operations.

A number of other stations, smaller in scope and known as naval auxiliary air stations, as well as naval reserve aviation facilities, were constructed to support the growing number of naval aviators. These additional stations were often located above the coastlines, typically intended for aircraft that did not immediately depend on the shore base for flight operations. An example is NAS Atlanta, Georgia. Constructed on the original site of Camp Gordon, a World War I Army base, today it is Atlanta's DeKalb Peachtree Airport.

Following the United States' entrance into the war, other naval air stations came into service. One such facility was NAS New Orleans, Louisiana, in 1943. This air station was constructed on the south shore of Lake Pontchartrain within a year's time to accommodate the massive volume of naval aviation training along the Gulf Coast. NAS New Orleans and other stations like it also served squadrons traveling across the country preparing to be deployed on carriers heading overseas.

1943. Future president George H. W. Bush (*standing, second from right*) trained at NAS Fort Lauderdale, Florida. (Photo courtesy of the Emil Buehler Library, National Naval Aviation Museum, NAS Pensacola, FL)

CREATION OF THE U.S. ARMY AIR FORCES

Like its Navy counterpart, the Army Air Corps also accelerated operations in the Gulf South. Bases were planned for locations along the Gulf Coast prior to the United States entering the war, while existing bases—among them Randolph, Barksdale, and Maxwell—were expanded. Further capitalizing on the utility of Army aviation, the U.S. Army Air Forces (AAF) was created in June 1941, organizing combat squadrons and operations at Army airfields under a single command. (It should be noted that the U.S. Army Air Corps remained active as a training, transport, and combat command of the Army until 1947, when the U.S. Air Force was formally established.)

The Army Air Forces made use of U.S. home-front airfields to train both American and Allied pilots, especially British and French crews who were unable to train on their native soil. In 1941, the AAF established the Southeast Air Corps Training Center (SEATC) and the Eastern Flying Training Command (EFTC). The EFTC was headquartered at Maxwell Field, Alabama.

Cities throughout the nation, including those in the Gulf states, were asked to make their aviation facilities available for military use. New construction of Army air bases occurred frequently on land designated by municipalities or states for future civil airports. This construction would become a postwar inheritance at minimal cost for many of these communities. By war's end, military bases had virtually taken over the map of the United States, with a facility located in nearly every city.

For this surge of Army expansion, the Gulf Coast offered some fundamental advantages over other parts of the nation, not the least of which was access to ports removed from hostile ocean waters and a year-round cooperative climate. Keesler Army Airfield in Biloxi, Mississippi, was commissioned prior to the war in June 1941 as Army Air Station No. 8. The base was home to an Army aviation mechanics school and also supported basic training for new recruits. Tyndall Air Base near Panama City, Florida, was opened in December 1941 as Gunnery School No. 9. The facility quickly grew to serve as a crew training base supporting other nearby facilities with its practice ranges. One of the more famous students to go through training at Tyndall was Clark Gable. Tyndall also provided training for Allied nations, including French and Chinese airmen. Other sites of Army expansion include Columbus Army Air Field, commissioned near the Mississippi city of that name in January 1942 as a twin-engine advanced training facility, and Selman Field in Monroe, Louisiana, which was established to provide navigator training on land originally set aside by the municipal government for a civil airport.

1943. NAS Pensacola at Chevalier Field with an extensive variety of aircraft, including the SNV, SNJ, PBY-5A, Grumman G-21 Goose, Beechcraft SNB-1 Kansan, Stearman N2S, and Dauntless dive bombers. (Photo courtesy of the Emil Buehler Library, National Naval Aviation Museum, NAS Pensacola, FL)

Left: A landing signal officer provides guidance to student pilots. (U.S. Air Force photo)

Right: **1942.** A Maxwell Army Air Base instructor climbs into the cockpit of a T-6 Texan trainer in preparation for the day's lessons. (Photo courtesy of the Air University/Air Force Historical Research Center, Maxwell Air Force Base, AL)

Tyndall Air Base, Florida, specialized in gunnery training for air crews, carrying out target practice and mock air combat over the Gulf of Mexico. (U.S. Air Force photo)

A Consolidated-Vultee trainer in flight outside of San Antonio, Texas. (Photo courtesy of the Air University/Air Force Historical Research Center, Maxwell Air Force Base, AL)

The Barksdale Army Air Base aircraft ramp in 1943. (Photo courtesy of the Barksdale Global Power Museum, 2nd Bomb Wing)

1943. Aviation cadets assemble on the ramp beside training aircraft at Naval Air Auxiliary Station Chase Field, Beeville, Texas. (Photo courtesy of the Emil Buehler Library, National Naval Aviation Museum, NAS Pensacola, FL)

Top: African American airmen at Keesler Army Air Field, Mississippi. (U.S. Air Force photo)

Bottom: Tuskegee airman and pilot prepare a P-51 Mustang "Red Tail" for flight. (U.S. Air Force photo)

MOTON AIR FIELD, TUSKEGEE INSTITUTE, AND TUSKEGEE ARMY AIR FIELD, ALABAMA

Apart from the rapid technological advances initiated by President Roosevelt to increase U.S. military readiness, cultural changes took place in the Gulf Coast region, many destined to impact the makeup of the U.S. military. One of the most important changes took place in 1939 in Tuskegee, Alabama, with the opening of the first African American primary flight school at Tuskegee Institute and the first African American advanced training programs at Tuskegee Air Field.

The founding of the aviation program at Tuskegee Institute was a milestone for the segregated U.S. military, the first step in providing equivalent aviation training for qualified African American pilot candidates. The program was conceived in response to the Civilian Pilot Training Program (CPTP), begun by the Roosevelt administration in 1938. CPTP-hosting institutions—designated colleges, universities, and flight schools at metropolitan airports across the country—restricted admission, and African American soldiers and citizens were not allowed to enter these programs. But because the number of trained pilots and crews might be inadequate if the United States was drawn into the war, Tuskegee was commissioned into primary aviation training. In turn, the institute was determined to demonstrate capacity for the demands of aviation.

Moton Air Field was the orientation center. Following completion of basic flight instruction, the pilots moved to nearby Tuskegee Army Air Field to begin advanced instruction. Each pilot inducted into the Air Corps was routed to fighter or bomber training. Members of these squadrons served in all war theaters and mastered piloting, navigation, and other skills. This training began at Tuskegee and was completed at bases in a variety of locations throughout the United States. Mechanics and squadron support personnel were also prepared at Tuskegee for advanced schools. In contrast to its unwarranted perception, the Gulf South thus became the birthplace of preliminary steps taken to desegregate the military.

The story of the Tuskegee Airmen has been romanticized in a number of ways, some of which have appropriately introduced this historic training program and its servicemen to new generations, while others retell the history with less accuracy. What is frequently overshadowed in appreciating the achievements of the Tuskegee Airmen is the fact that the program not only trained fighter pilots but also—throughout the war and for some time after—produced graduates for a variety of other advanced flight assignments and duties at bases throughout the United States, leaving no doubt of the capacity of African American personnel to fulfill such critical responsibilities.

The North American P-51 Mustang "Red Tail" has become the popular icon of the Tuskegee Airmen. A number of other aircraft, equally demanding of skill, were also flown by Tuskegee-trained pilots, including the Curtiss P-40 Warhawk and the North American P-47 Thunderbolt fighters. Tuskegee bomber squadrons flew Beechcraft AT-10 multiengine trainers and the North American B-25 Mitchell, the aircraft model used by Gen. James Doolittle in the 1942 raid on Tokyo.

THE SUBMARINE WAR IN THE GULF

Not long after the Germans declared war on the United States, Nazi U-boats began stalking vessels in the Gulf of Mexico, preying on American shipping within sight of the coastline. The Gulf soon became one of the most threatened areas of the home front. The Gulf ports and the mouth of the Mississippi were irresistible targets monitored by German U-boats throughout the war. Affecting its reputation as a shipping safe haven, by early 1942, waters from the Atlantic coast of Florida across the Gulf to southernmost Texas were under continuous observation by Nazi submariners.

As a result of this enemy activity, bases throughout the Gulf Coast states added active home-front defense to its mission of training aviators. A series of newly constructed air bases supported a variety of such operations across the shoreline. Naval bases used slower patrol planes and airships as submarine spotters. Air Corps pilots also patrolled in search of the enemy presence. Although a number of Allied vessels were destroyed by German U-boats in the Gulf of Mexico, records indicate that only a single U-boat, the U-166, was sunk by U.S. coast patrols of the Gulf.

POST–WORLD WAR II SURPLUS BASES

The approaching end of World War II presented a new problem amid the spoils of victory. There was an extreme surplus of military material, equipment, and real estate throughout the defense network that was now an unnecessary burden on service operations and budgets. An examination of options ultimately resulted in the Federal Property and Administrative Services Act of 1949, created to efficiently distribute the property and infrastructure no longer of use to the armed services. Since then, there has been an ongoing closure and consolidation of U.S. military bases around the globe.

During the war, the home front changed dramatically in appearance. Military bases were rapidly constructed, often only miles apart from one another, resulting in a nation covered in a network of facilities prepared to support any number of operations. Much of the duplication among bases and within regions was necessary for security, in the event a similar base was incapacitated. Following the war, many of these facilities were consolidated or completely decommissioned. Some closures took place during the last year of the war before the end of hostilities. Harding Army Air Field in Baton Rouge, Louisiana, for example, rapidly reduced active military operations and began a gradual transition to a full-service civilian airport in preparation for postwar use by the community.

At the outbreak of World War II, NAAS Houma, Louisiana, was rapidly constructed on the site of a civilian airport. The facility featured a runway and airship hangar to support patrol operations in the Gulf of Mexico. (Photo courtesy of the Houma–Terrebonne Airport/Industrial Park)

April 1942. Lt. Col. Jimmy Doolittle, with co-pilot Col. Dick Cole, fly the first B-25 bomber from the USS *Hornet* for the strike by the United States on the homeland of Japan. (Photo courtesy of the U.S. Navy)

THE DOOLITTLE RAIDERS

Following the devastating attack on Pearl Harbor in December 1941, the U.S. military leadership quickly looked for ways to take the fight back to Japan. James Doolittle, known for his air-racing victories in the early 1930s and possessing an encyclopedic knowledge of aviation operations, agreed to lead sixteen crews flying North American B-25 medium bombers from the aircraft carrier USS *Hornet*, positioned off the coast of Japan, on a sneak attack designed to shake that nation's confidence and boost American morale. The *Hornet* was the Navy's newest carrier, commissioned only six months earlier in October 1941, and was destined to sail for only a single year, being sunk at the Battle of Santa Cruz in October 1942.

A significant Gulf Coast element was critical to the mission's success. The Raiders' training was conducted at Eglin Army Air Field, located between Shalimar and Valparaiso, Florida, as well as on Wagner Field, Eglin Auxiliary Field No. 1, in the deepest secrecy. Navy flight instructors from NAS Pensacola prepared a training program and taught these volunteer Air Corps pilots how to take off from the deck of an aircraft carrier. At the time, the B-25 was the largest aircraft to operate from a naval vessel for any nation. The aircraft were modified, reducing the operating weight by the removal of any unnecessary on-board equipment. Upon completion of training at Eglin, the crews flew the aircraft—now famously "at treetop level"—across the United States to San Francisco, where they were loaded onto the waiting *Hornet*.

The mission, commonly known as the Doolittle Raid on Tokyo, flew on April 18, 1942, and caught the Japanese completely off guard. The psychological damage was enormous, shocking the enemy into a lasting sense of insecurity. Every aircraft escaped the Japanese mainland. Although not one of the aircraft was shot down, a number of crew members were captured. Running out of fuel over China, the crews were forced to parachute from the aircraft or crash land. Survivors had to find their way to friendly assistance that would give safe harbor until they could be reunited with U.S. forces. One aircraft landed in Vladivostok, Russia, where the crew were held for the duration of the war. Many who made it back to the United States fought on through the war's end in 1945.

Eglin Air Force Base is now honorably referred to as the home of the Doolittle Raiders. Their "Farewell Tribute" 71st anniversary reunion took place on the base April 17–20, 2013. Three of the four surviving members of the original group of eighty Raiders were able to attend that last official gathering at Eglin, closing a unique chapter in Gulf Coast military history.

2013. Three of the four surviving Doolittle Raiders meet at Eglin Air Force Base, the site of their training for the mission. From left, Col. Dick Cole, Lt. Col. Ed Saylor, and Staff Sgt. David Thatcher salute during the National Anthem at the 33rd Fighter Wing hangar dedication in honor of Saylor. Surviving Raider Lt. Col. Robert Hite was unable to attend the ceremony. (U.S. Air Force photo/Samuel King Jr.)

The surviving Raiders agreed to assemble one last time on November 9, 2013, at the United States Air Force Museum at Wright-Patterson Air Force Base in Dayton, Ohio, where a ceremony was held to incorporate the close of their mission into the museum display. Deep in tradition, the squadron mates modified a longstanding plan for sharing a bottle of Cognac originally reserved for the last two survivors. The three attending squadron members offered a final toast to their departed, then handed over to the museum the bottle of Cognac and goblets used in the toast, for a permanent display.

On May 23, 2014, the members of the Doolittle Raiders, officially known as the 17th Bombardment Group, were awarded the Congressional Gold Medal, the "highest expression of national appreciation for distinguished achievements and contributions" bestowed by the U.S. Congress.

1941. In spite of its inability to take off or land on surfaced runways, the PBY-5 proved to be one of the most versatile aircraft of World War II, retrieving thousands of downed pilots and conducting long-range patrols other aircraft were incapable of completing. (Photo courtesy of the U.S. Navy)

THE CONSOLIDATED PBY CATALINA FLYING BOAT

Consolidated Aircraft Company was founded in 1923 by Reuben H. Fleet in Buffalo, New York. In 1935, Fleet relocated the company to San Diego, California, for access to open water and a more favorable year-round climate. The following year, the U.S. Navy ordered Fleet's design for a new flying boat into production, designating the aircraft "PBY" for Patrol Bomber class (the Y identified Consolidated Aircraft Company as the manufacturer).

Fleet's model was revolutionary, introducing a much larger wing surface area (over twice that of previous aircraft) positioned above a completely enclosed aerodynamic boat hull fuselage. The result was a long range and a far greater lifting capacity for passengers, armament, and cargo. By 1941 when the U.S. entered the war, the PBY had become a standard of design for U.S. Navy flying boat operations and was serving Allied nations in the European and Pacific theaters.

Competing aircraft manufacturers proposed that the PBY was too slow in a rapidly changing aviation world. However, its unique asset of range and stability at very slow speeds made the aircraft ideal for long-range patrols and rescue over Atlantic and Pacific waters. The name Catalina came from British pilots flying the craft in the early years of World War II. Success resulted in the PBY-5A and PBY-6A amphibian models.

In 1943, Fleet opened Consolidated Aircraft Company–New Orleans to produce the PBY amphibians in the numbers ordered by the U.S. Navy and allied nations. Located at the end of Franklin Avenue on the shores of Lake Pontchartrain, the plant in the course of a little over two years employed more than five thousand personnel and produced approximately three hundred PBY aircraft. A large number of the aircraft manufactured in New Orleans were exported to the Allies.

Completed aircraft were moved from assembly line to Lake Pontchartrain for final testing and delivery. There was no runway at the facility, only a ramp leading to the lake waters. Each PBY constructed at the New Orleans plant was a true amphibian, meaning the aircraft had retractable on-board landing gear, unlike its pioneering PBY predecessor models constructed at the San Diego plant, which were wheel-less flying boats.

Production of the legendary PBY aircraft ended at the New Orleans assembly line in September 1945 when the last aircraft, a model 6A, was delivered. Today, aircraft museums across Europe and Asia feature in their static display collections PBYs built at the New Orleans factory between 1943 and 1945. A number remain airworthy worldwide. The growing popularity of restoring these aircraft may increase this number in the future.

Plate 1. After Hurricane Ivan made landfall between Gulf Shores, Alabama, and Pensacola in 2004, 90 percent of NAS Pensacola's buildings experienced damage. (Photo courtesy of the Public Affairs Office, NAS Pensacola)

Plate 2. 1938. A Grumman F3F-2 fighter assigned to the aircraft carrier USS *Yorktown* (CV-5) flies over the coastline. In 1936, the Navy began incorporating the F3F-1 into carrier- and land-based squadrons. (Photo courtesy of the Emil Buehler Library, National Naval Aviation Museum, NAS Pensacola, FL)

Plate 3. 2011. (Lt. Col.) Dr. Granville C. Coggs and (Sgt.) Warren Eusan, original Tuskegee Airmen, address the audience at the Third Annual Tuskegee Heritage Breakfast, held at the 99th Flying Training Squadron on Randolph Air Force Base. (U.S. Air Force photo/Rich McFadden)

Plate 4. 1943. Throughout World War II, NAS Corpus Christi and NAS Jacksonville trained thousands of naval aviators in the PBY Catalina flying boat, using ramps for the original non-amphibious models. Later versions, the PBY-5A and 6A, were land based through the addition of on-board landing gear. (Photo courtesy of the U.S. Navy)

Plate 5. 2003. A T-45 Goshawk from NAS Kingsville, Texas, lands on the flight deck aboard the *John C. Stennis*. T-45s are used for intermediate and advanced portions of the Navy and Marine Corps pilot training programs for jet carrier aviation and tactical strike mission instruction. (U.S. Navy photo/Photographer's Mate 3rd Class Joshua Word)

Plate 6. Air National Guard units serving across the Gulf South, such as the 136th Airlift Wing based at NAS Fort Worth, are vital not only to homeland security and emergency response but also to the overall global defense network. (U.S. Air Force photo)

Plate 7. An Alabama Air National Guard KC-135 positions over the Gulf of Mexico to refuel F-15s and an F-22 Raptor from Tyndall AFB. (Photo courtesy of the U.S. Air Force)

Plate 8. A CV-22 Osprey from the 8th Special Operations Squadron and an MH-53 Pave Low from the 20th Special Operations Squadron fly over the coastline near Hurlburt Field. The CV-22 replaced the MH-53 in the Air Force Special Operations Command. (U.S. Air Force photo/ Senior Airman Julianne Showalter)

Plate 9. An AC-130U gunship from the 4th Special Operations Squadron jettisons flares over an area near Hurlburt Field. The flares are a countermeasure for heat-seeking missiles that may be launched at the aircraft during real world missions. (U.S. Air Force photo/Senior Airman Julianne Showalter)

Plate 10. 2009. Strike Fighter Squadron VFA-204 F/A-18 Hornets and Louisiana Air National Guard 159th Fighter Wing F-15 Strike Eagles fly in an echelon formation over the Mississippi River and south Louisiana wetlands. (U.S. Navy photo/Mass Comm. Specialist 2nd Class John P. Curtis)

Plate 11. 2014. Four F-5 Tiger fighters from the VFC-111 Sundowners on the flight line at NAS Key West's Boca Chica Field. The Sundowners provide realistic combat training over the Gulf of Mexico to fighter and attack pilots from all U.S. military branches and numerous allies. (U.S. Navy photo/Mass Comm. Specialist 2nd Class Brian Morales)

Plate 12. A Canadian F-18 from the 425th Tactical Fighter Squadron, Bagotville, Quebec, and other squadron aircraft fly over Fort Jefferson and the waters of south Florida. Homestead Air Reserve Base, Florida, also hosts allied squadrons participating in Gulf air-to-air exercises. (Photo courtesy of the U.S. Air Force)

Plate 13, *top*. A Navy TA-4 Skyhawk, T-2 Buckeye, and T-45 Goshawk from NAS Meridian fly together above central Mississippi. (U.S. Navy photo)

Plate 14, *bottom*. The T-38 Talon, a twin-engine, high-altitude, supersonic jet trainer, flies high over north Mississippi near Columbus AFB. (Photo courtesy of the U.S. Air Force)

Plate 15. This Beechcraft T-34 Mentor trainer (*center and top*) at NAS Whiting Field, Florida, was painted in a retro 1930s livery to celebrate the one hundredth anniversary of naval aviation in 2011 (see plate 2). (Photo courtesy of the U.S. Navy)

Plate 16. 2010. The Blue Angels at NAS Joint Reserve Base New Orleans, Louisiana. Blue Angel #7 is performing as one of the four diamond aircraft, providing a VIP demonstration flight in the formation's "Echelon Parade" maneuver. (U.S. Navy photo)

Plate 17. March 27, 2008. Four T-38s from the 560th Flying Training Squadron, assigned to the 12th Flying Training Wing, fly over the historic district of Randolph Field. Three retired Air Force pilots who were POWs in Vietnam flew backseat during the mission as part of Randolph's ceremonies to honor their sacrifices. (U.S. Air Force photo/Master Sgt. Scott Reed)

CHAPTER FOUR

KOREA, VIETNAM, AND THE COLD WAR

The 38th Parallel is the recognized boundary between North Korea and South Korea. Slicing the Asian nation in half, it is a precise divide resulting from Japanese forces caught below the 38th surrendering to the Americans, and those above it surrendering to the Soviet Union at the end of World War II. On June 25, 1950, the forces of North Korea overran the parallel, with the objective of reunifying the country as a Communist nation. The resulting Korean War was the first aggressive action between western democratic and Communist military forces. In the words of former British prime minister Sir Winston Churchill, an "Iron Curtain" separated the victors of World War II, initiating a complex array of alliances and pacts. American foreign policy now focused on the concept of "containment" of communism.

In the second half of the twentieth century, air stations and bases in the Gulf Coast states trained personnel and supported aircraft squadrons operating in threatened regions of the world. At the same time, a new era in military aviation began with jet-powered wings. In addition, rapid cultural shifts and civil rights legislation placed the Gulf South at the center of some of the most volatile events in U.S. history. Service personnel found themselves in an uncharted middle ground that lay between politics, their own youth, and the combat zones. Military aviation operations within these states, often unbeknownst to the neighboring public, were also a hotbed of innovation destined to influence generations of flyers to come.

For the purpose of perspective, if World War II was an era of exponential expansion of U.S. military strength, then the period from the Cold War to the end of the twentieth century was, with few exceptions, one of consolidation and cost control. Uniform aircraft and weapon arsenals for all branches of the armed services created a range of operational adaptations intended to streamline the nation's massive military budget. Based on a new universal utility, aircraft were ordered from manufacturers under the direction of the Department of Defense as much for their potential to support all services as to fill the needs of a single branch. The outcome of this approach has been debated ever since, and its impact felt throughout the military installations of the Gulf South.

DEMOBILIZATION

In the years between 1945 and 1955, and then beyond, a unique mix of reductions and modernization took place on military bases spanning from Texas to Florida in anticipation of a new coexistence with communism. Pilots returned to their civilian lives, and a greatly reduced number of active personnel were reorganized for a peacetime

Top: One of the most important missions of the new U.S. Air Force was air transportation of personnel and materials overseas. During World War II, this task was accomplished almost entirely by sea. From 1950 through the mid-1970s, the C-152 Globemaster was a primary provider for this evolving assignment. (Photo courtesy of the U.S. Air Force)

Bottom: The Korean War was the first conflict to utilize U.S. jet aircraft in combat. The transition included state Air National Guard units. Here, a ground crew prepares an F-84 Thunderjet from the Florida ANG 159th Fighter Squadron for a combat mission from a base in South Korea. (Photo courtesy of the Florida Air National Guard/U.S. Air Force)

military. In 1945, the Navy deployed twenty-eight aircraft carriers. By June 1950, eleven remained, with even fewer available for active duty. Upon Japan's surrender in September 1945, the U.S. Army Air Forces had nearly 80,000 aircraft in inventory. By December of that year, an accelerated program for dramatically reducing the number of operating aircraft was initiated, and would last through the next decade.

The expense of returning equipment from overseas weighed against the threat of its falling into hostile hands often resulted in aircraft and support ground vehicles being destroyed in the Pacific and Europe. At home, millions of tons of equipment were disposed of or stripped of usable parts and the remaining metal prepared for sale in bulk. Mainland military facilities were in a variety of conditions. Much like those at the end of World War I, many had been intended for temporary rather than ongoing permanent use and thus posed another complication in postwar demobilization.

THE BIRTH OF THE U.S. AIR FORCE AND THE BEGINNINGS OF JET AVIATION

In 1947 the United States Air Force was created as an independent military branch, separate from the Army. This official divide of Army air and ground forces had been debated as far back as World War I, but the years following the surrender of Germany and Japan afforded the opportunity to execute it without compromising the service's capacity to respond to threats. The decision to follow through with the separation was simple. By the end of World War II, aviation was no longer a support force for ground troops and artillery. It was now an entity providing its own offensive and defensive missions with and without the use of accompanying ground forces.

The new service grew into a network of commands and squadrons with specific missions. In the Gulf South, the U.S. Air Force assumed command of former AAF bases, among them Barksdale, Columbus, Eglin, Keesler, Maxwell, Moody, and Randolph. Some personnel at these facilities were reassigned to new units defined to support aviation, while others continued to support the Army ground forces through an air transport service that would evolve into the Military Airlift Command. Air Force personnel received new uniforms distinguishing them from their Army counterparts in comparative rank. The newly commissioned U.S. Air Force lengthened runways and expanded service facilities of select former AAF bases across the Gulf South.

Left: **1950.** The Northrop F-89C, seen here landing at Eglin Air Force Base, was the first all-weather fighter-interceptor delivered to the U.S. Air Force that was designed to carry nuclear weapons. (U.S. Air Force photo)

Right: While escorting B-29s near "MiG Alley" over North Korea on June 26, 1951, 1st Lt. Arthur E. Olinger and Capt. Harry Underwood of the 182nd Fighter Bomber Squadron, Texas, shared credit for the Air Guard's first jet kill, a MiG-15. (Photo courtesy of the Texas Air National Guard/U.S. Air Force)

The technology for jet aircraft was available during the latter years of World War II but developed too late for all but limited practical use in the theaters. In 1944, Germany began operating a successful jet aircraft, the Messerschmitt ME-262, with effectiveness in defense against Allied bomber raids over its homeland. The British also operated a lesser-known jet aircraft in 1944 and 1945 named the Meteor, which first saw action in pursuit of Nazi V-1 bombs traveling over the English Channel en route to targets in England. The arrival of jet aircraft in the United States brought a new performance standard. Even the earliest experimental models offered advantages, among the most beneficial being speed and higher operating altitudes. Powered by turbine- rather than piston-driven engines, these aircraft were incorporated into squadrons at Air Force bases and naval air stations. New jet squadrons demanded training programs for pilots, mechanics, line technicians, and other support personnel. The warm Gulf South climate and traditional location of aviation training attracted construction of a number of new schools and squadron transition programs.

The introduction of jet aircraft reduced the utility of Chevalier Field, NAS Pensacola. The T2V Sea Star was the last jet to operate from the aging airfield. (Photo courtesy of the Emil Buehler Library, National Naval Aviation Museum, NAS Pensacola, FL)

Sherman Field, NAS Pensacola, was constructed on the western side of the air station. This mid-1980s photograph shows the parallel east–west runways accompanied by a single crosswind strip. (Photo courtesy of the Emil Buehler Library, National Naval Aviation Museum, NAS Pensacola, FL)

Top: This photograph reflects the enormous difference in size between the World War II–era B-29 Stratofortress and its Cold War replacement, the B-36 Peacemaker, both once based at Carswell AFB, Texas. Design of the B-36 began before the end of World War II and was continuously modified as the necessity of vast global range and nuclear payload became increasingly evident. (Photo courtesy of the U.S. Air Force)

Bottom: The B-58 Hustler was manufactured by Convair in Fort Worth, Texas, and a bomber group was for a time based at Carswell AFB. (U.S. Air Force photo)

THE ATOMIC DILEMMA

The advent of the atomic bomb increased demand for larger aircraft capable of delivering the new weapons. In 1948, the recently commissioned Air Force proposed to develop the Convair B-36 Peacemaker. The B-36 was designed in the final years of World War II, and the Peacemaker was to be the world's largest long-range heavy bomber as well as the largest propeller-driven aircraft ever built. It would fly around the world from, and return to, the United States. A number of Gulf States bases would be enlarged to support aircraft squadrons responsible for B-36 operations. Barksdale AFB, Louisiana, and Carswell AFB, Texas, among other facilities, were to host the flight crews and aircraft charged with that mission.

That same year, the Navy proposed constructing a super aircraft carrier, the USS *United States*, capable of delivering strike aircraft and nuclear weapons across the globe from the seas. While the carrier itself would likely be based in a region outside of the Gulf Coast, training and other necessary infrastructure for the vessel—as well as for other future carriers—would involve Gulf Coast facilities.

At stake was exorbitant funding, which caused a great rift between the Air Force and the Navy. Questions from both military branches quickly followed. Would Air Force aircraft, Navy aircraft, or both be best suited to deliver these weapons in case of need? Admirals and generals understood that decisions made would mean a paradigm shift in service mission. The military branch funded would become the dominant force in U.S. defense operations, while those with fewer dollars committed to their expansion would likely play more of a supporting role in any strategic actions.

The USS *United States* program was approved by President Harry S Truman in 1948 but subsequently canceled that same year in favor of the B-36 project. The aircraft was built and deployed in 1948 to, among other bases, several within the Gulf states, including Carswell. It replaced the B-50 as the principal bomber in the Air Force arsenal. Within the next decade, though, another shift in plan occurred, this time to replace the B-36 itself with newly developed jet bombers, the B-46 and B-52, models that had quickly surpassed the utility of the B-36.

By the mid-1950s, as the Cold War accelerated, NAS Sanford, Florida, replaced the aging PB4Y-1 patrol bombers of World War II (*center*) with the North American AJ Savage (*left and right*), allowing for increased payload and the capacity to carry nuclear weapons. (Photo courtesy of the Emil Buehler Library, National Naval Aviation Museum, NAS Pensacola, FL)

MILITARY CONSOLIDATION AND ITS IMPACT ON AVIATION IN THE GULF SOUTH

The Department of Defense was established in 1949 by an amendment of the National Security Act and placed under the direction of a cabinet-level secretary. The role of the secretary of defense evolved throughout the 1950s, and one of the first substantive tests of the position's effectiveness in leading joint military planning came as a byproduct of the Vietnam War. In the early 1960s, under Secretary of Defense Robert McNamara in the Kennedy administration, the trend toward multi-branch use of military assets developed strength, and aircraft design and operations were among the first to be affected.

Several aircraft stationed on or transiting through bases in the Gulf South were precedent-setting success stories in this objective. The F-4 Phantom II was originally flown in 1958 by McDonnell Aircraft Co. to be a carrier-based fighter for the Navy. Unlike previous fighters intended for a single purpose, the Phantom was modified throughout its life and served as a fighter and bomber with the Air Force and Marines. Other cross-service successes during the Vietnam War include the Chance Vought A-7 Corsair and the Lockheed C-130 Hercules, which the Air Force, Marines, Navy, and Coast Guard, as well as numerous allied nations, incorporated into service for their own purposes.

Some models, successful in a specific design and mission, were historic failures of the secretary's vision. For example, the General Dynamics F-111 Aardvark was originally intended to be both a carrier-based fighter for the Navy and a land-based multiple-use aircraft for the Air Force. Following years of debate, the Navy withdrew its interest. The aircraft continued to be developed and entered service with the U.S. Air Force in 1967.

The Base Realignment and Closure Act of 1990 formally continued the policy that began under the Kennedy administration. Its objective was the ongoing examination of military bases in close proximity, both geographically and in terms of mission. Following extensive study, the secretary of defense provided recommendations to the Base Closure and Realignment Commission, which in turn presented them to Congress for determining the fate of these federal properties. Communities throughout the nation, including the Gulf South states, were well aware that the closing of bases would negatively impact local economies.

The naval air and Coast Guard stations, Air Force bases, Army forts, and Marine camps considered essential assumed the capital assets, flight squadrons, and duties of those bases scheduled for closure. A number of air squadrons created to support specific operations in Korea or Vietnam were disbanded, similar to the multitude of squadrons decommissioned after World War II in the transition to peacetime operations. According to U.S. Air Force statistics, "Between 1946 and the first decade of the 21st Century, 90% of the major military and auxiliary airfields constructed during World War II were inactivated." The total exceeds five hundred. Throughout the Gulf South, a number of these bases were transferred to municipalities for use as new civil airports or in land development. The majority of these locations were transformed, with the exception of small facilities remaining on-site adapted for reserve or National Guard personnel acclimating to an active or inactive status within these communities.

THE CUBAN MISSILE CRISIS

Possibly the most intensive test for military aviation on the Gulf Coast was the Cuban Missile Crisis in October 1962. The island nation of Cuba and its leader, Fidel Castro, posed the first real threat of nuclear war and missile attacks on the United States mainland. Air Force bases along the Gulf Coast and Navy aircraft carriers stationed in the Gulf launched photographic reconnaissance missions to maintain constant surveillance of the Caribbean island, while every military facility from Florida to Texas prepared to support air strikes should it become necessary. Naval Air Station Key West, Florida, the closest to Cuba (less than one hundred miles from the island), supported a naval blockade ordered by President John F. Kennedy, with reconnaissance flights throughout the crisis. U-2 "spy planes" operating from McCoy Air Force Base (today's Orlando International Airport) flew daily over Cuba to conduct photo reconnaissance.

Residents of the Gulf Coast witnessed the United States' response to the Soviet-built weapons then in fatal proximity. Railroad cars transported armored equipment in preparation for an assault on Cuba. Convoys overwhelmed Gulf Coast highways, moving troops and equipment from base to base. Even civil airports throughout the Gulf states were on alert as transports brought thousands of soldiers to Army forts in the Gulf Coast states. By the first week of November 1962, the crisis was resolved. The Soviet Union agreed to remove the missiles with assurance from the United States that it would not invade the island once the weapons had been withdrawn.

Above: Numerous aircraft, including the F-101 Voodoo reconnaissance fighter, flew over Cuba from bases along the Gulf Coast to capture images of the daily activity during the October 1962 Cuban Missile Crisis. (U.S. Air Force photo)

Left: The Boeing B-47 Stratojet, predecessor to the B-52, was stationed at, among other airfields, Chennault AFB in Lake Charles, Louisiana, during the Cuban Missile Crisis. (U.S. Air Force photo)

1971. For a time during the Vietnam War, it was thought that the F-8 Crusader might be the last U.S. fighter to be equipped with on-board guns. This VF-202 Crusader based at NAS Dallas, Texas, flew aboard the USS *John F. Kennedy.* (Photo courtesy of the Emil Buehler Library, National Naval Aviation Museum, NAS Pensacola, FL)

The F-4 Phantom II became a symbol of fighter strength for Air Force, Navy, and Marine units in Vietnam. Many Phantoms were home based along the Gulf Coast. (Photo courtesy of the U.S. Air Force)

THE HELICOPTER: VERTICAL FLIGHT AND ROTARY WING AVIATION

Although in the latter years of World War II, the Army, Marines, Navy, and Coast Guard experimented with the helicopter—rotary wing aviation—practical vertical flight came to fruition following the separation of the Air Force into an independent service. As with the airplane, mission-specific needs such as search, rescue, and forward air control inspired a number of diverse designs. All of the branches, but especially the Army and Marine Corps, promptly discovered the helicopter's utility in battle, adapting it for transportation of weapons, supplies, and personnel on short and medium distances to combat areas.

By the late 1950s, the Army's massive incorporation of the helicopter transformed Alabama with the construction of principal operating centers in the state. Primary helicopter flight instruction for all of the branches as well as some advanced squadron-specific crew training was conducted at training facilities already operating in the Gulf South states. A new round of service rivalry ensued, especially between the Army and Air Force, concerning who could best command helicopter operations, with each branch soon discovering its own unique specialties for vertical flight.

Early use of the Army helicopter was portrayed in the television program *M*A*S*H* (Mobile Army Surgical Hospital), which depicts medical units during the Korean War. In that same era, Navy helicopters began retrieving pilots from open-water accidents during aircraft carrier launch and recovery operations, and select Navy vessels were equipped with deck helicopter landing pads for sea operations. The 1960s introduced a variety of helicopter models. Flying in the skies of Southeast Asia during the Vietnam War—and among the most photographed—were the Army HU-1 Iroquois and the Boeing Vertol CH-47 Chinook, with its unique twin blades forward and rear. Both were used for moving personnel, cargo, and artillery to the front lines. The Navy employed the helicopter for anti-submarine patrol, first outfitting larger SH-3 Sea King and later MH-60 Sea Hawk models for global anti-submarine defense, using the Gulf of Mexico as both a training and an operating area.

More than any other conflict in aviation's brief history, the Vietnam War demonstrated and proved the utility of the helicopter. Its use over jungle battlefields, few of which allowed room for landing strips—even the smallest envisioned—brought about an entirely new discipline in military aviation. Rotary-wing operations throughout the Vietnam conflict provided the foundation for Army and Marine direct insertion of troops and equipment onto front-line combat areas.

Historians mark 1990 as the end of the Cold War between the United States and the Soviet Union, following forty-five years of mutual international mistrust. During this period, with considerable credit due to operations in the Gulf states, the helicopter developed into an aircraft with great versatility. Uses ranged from Army and Marine attack and reconnaissance battalions to heavy-lift utility and rescue squadrons operated by Navy, Air Force, and Coast Guard commands.

Top left: The EC-121 Warning Star was used by the Air Force and Navy to provide radar and the earliest electronic warfare operations during the Vietnam conflict. Many crew members trained at Gulf Coast bases and used an array of on-board equipment to provide pilots and ground personnel with intelligence on Viet Cong troops and aircraft. (National Museum of the Air Force website; photo courtesy of the U.S. Air Force)

Top right: Throughout the Vietnam War, the U.S. Navy A-3 Skywarrior was used extensively as an electronic countermeasure aircraft. (U.S. Navy photo)

Bottom: **1967.** An F-100 Super Sabre fighter of the 90th Tactical Fighter Squadron at Bien Hoa Air Base, South Vietnam. Established as the 90th Aero Squadron at Kelly Field, Texas, in 1917, to date this squadron remains one of the oldest standing lineages in the U.S. Air Force. (U.S. Air Force photo)

Top: The Huey UH-1 helicopter was the workhorse of the Vietnam war, its utility reminiscent of the DC-3 from World War II. (Photo courtesy of the U.S. Air Force)

Bottom: **1968.** An Air Force combat rescue team designed to support the HH-53 "Jolly Green Giant" included A-1 Skyraider strike aircraft and a C-130 Hercules aerial refueler. Rescue training was and continues to be conducted at Dobbins AFB and Moody AFB, Georgia. (U.S. Air Force photo)

Right: NAS Meridian (McCain Field), Mississippi, opened in 1961 operating the T-2 Buckeye advanced jet trainer. (Photo courtesy of the U.S. Navy/Public Affairs Office, NAS Meridian)

Below: NAS New Iberia, located in the heart of Louisiana's Cajun country, operated from 1960 through 1965 as a training base for multi-engine aircraft such as the S-2 Tracker. (Photo courtesy of the U.S. Navy)

NEW NAVAL AIR STATIONS

In the wake of base closures and consolidations, the Navy repositioned assets for its advanced training aviation program, opening new bases in the Gulf South. In central Mississippi, NAS Meridian opened in July 1961 to support the increased need for carrier-qualified naval aviators. The base provided primary and advanced instruction in the newest jet trainer, the North American T-2 Buckeye. Its field was named in honor of Mississippi native Admiral John McCain, Sr., the first of four generations of the McCain family to graduate from the U.S. Naval Academy. Admiral McCain served with distinction through the end of World War II, dying of a heart attack on September 6, 1945, only four days after attending the Japanese surrender on the USS *Missouri* in Tokyo Bay. His son, Admiral John McCain, Jr., is the father of Senator John McCain of Arizona, a decorated naval aviator. To this day, NAS Meridian, McCain Field, continues to offer intermediate and advanced jet aircraft training for naval aviators who have completed primary and joint intermediate flight instruction at stations throughout the Gulf Coast. NAS Meridian is a station for the McDonnell Douglas (Boeing) T-45 Goshawk jet trainer, which introduces naval aviators to their first fleet aircraft carrier landings.

Naval Auxiliary Air Station New Iberia, Louisiana, opened in 1960 to support training of naval aviators flying the Grumman S-2 Tracker, E-1 Tracer, and the A-1 Skyraider. Graduates of NAAS New Iberia were assigned to carrier-based early warning squadrons. However, in 1965, after only five years of operation, the station was closed. Its assets and squadrons transferred to NAS Corpus Christi.

THE HUEY: SUPREME TRANSPORT VEHICLE OF THE VIETNAM WAR

One of the most iconic images of the Vietnam War is the helicopter, specifically the HU-1 Iroquois, better known as the "Huey." This aircraft became the supreme field transport vehicle, no less critical to personnel in Southeast Asia than the rugged jeeps, half-tracks, and Douglas C-47 (DC-3) aircraft were to the logistical needs of World War II. The Huey was associated with nearly every mission and battleground during the conflict, from training to front-line combat, transporting troops into and out of enemy gunfire.

Designed and manufactured by Bell Helicopters, now headquartered in Fort Worth, Texas, the Huey was the single most recognizable vehicle that changed the reputation of the helicopter from one of questionable usefulness to that of the most appreciated and practical craft in the war. Although helicopters appeared by the end of World War II and were used in the Korean War, it was the image of the Huey in action in Vietnam that brought rotary wing military flight to public attention.

The aircraft's original designation of HU-1 (Helicopter Utility Model 1) evolved into the now-famous nickname Huey. The Iroquois, though taking a backseat to the popular Huey, has made a comeback as modifications to the original aircraft continue in the twenty-first century. Today, the Huey is referred to in the uniform military nomenclature as UH-1, a simple reversal of letters that conforms to aviation equipment classifications throughout the Department of Defense and among its primary users.

Versions of the helicopter were also used extensively by the Navy, Air Force, and National Guard units for a variety of assignments. One of the more visible was medical service, evacuation, and rescue. Like the B-52 bomber, some aircraft designs stand the ravages of time far beyond their contemporaries. With avionics and engine upgrades and modifications, the UH-1 airframe design and the surviving fleet are expected to provide reliable service through the mid-twenty-first century.

Top: Students piloting UH-1H Hueys from the 23rd Flying Training Squadron practice formation flight over the runway at Fort Rucker, Alabama. (U.S. Air Force photo/Airman 1st Class Anthony Jennings)

Bottom: To this day, the Bell UH-1 lineage continues to play a vital role in the Air Force inventory. Here, UH-1Ns operate at Duke Field, Eglin AFB. (U.S. Air Force photo/Samuel King Jr.)

For over four decades, the USS *Lexington* conducted carrier qualifications in the Gulf of Mexico for naval aviators training at each of the southern naval air stations. (U.S. Navy photo)

Opposite: The USS *John C. Stennis* (CVN-74) is one of three active modern-day aircraft carriers not named after a U.S. president. Christened in honor of Mississippi's energetic senator, the massive nuclear-powered ship is seen here at sea conducting landing qualifications for student aviators. (Photo courtesy of the U.S. Navy)

AIRCRAFT CARRIER TRAINING OPERATIONS IN THE GULF OF MEXICO

Naval air stations on the Gulf Coast have had an association with aircraft carrier operations since the commission in 1922 of the first "flattop," the USS *Langley* (CV-1), which made trips to the Gulf Coast to support the earliest flight training. Before World War II began, the USS *Ranger* (CV-4) also sailed in the Gulf, providing training for student pilots. However, World War II training of naval aviators to take off and land on aircraft carriers was conducted in the safe harbor of the Great Lakes, using two converted passenger vessels, the USS *Wolverine* and USS *Sable* (steamers featuring a flat deck and uniquely retaining their side paddlewheels for propulsion). Although far removed geographically from the primary training conducted in Florida and Texas, the carriers avoided the U-boats actively patrolling the Gulf of Mexico and Atlantic shores. In the years immediately following Japan's surrender, smaller escort carriers, including the USS *Saipan* (CVL-48), USS *Wright* (CVL-49), USS *Cabot* (CVL-28), and USS *Monterey* (CVL-26), rotated into the Gulf to support flight training.

Cold War military aviation rapidly evolved from propeller-driven to larger turbine- and jet-propelled aircraft. As a result, the Navy required a larger carrier to handle the more powerful planes. In 1957, the USS *Antietam* (CVS-36), an upgraded and modified World War II *Essex*-class carrier, was permanently assigned to NAS Pensacola to support training for the Gulf South bases. Following the 1962 Cuban Missile Crisis, the responsibility was handed over to *Antietam*'s sister ship, the USS *Lexington* (CV-16), which had been deployed to the blockade of the Communist nation.

For three decades, the *Lexington* became the symbol of pilot carrier qualifications in the Gulf of Mexico. NAS Pensacola, Corpus Christi, and Meridian provided a continuous influx of student naval and Marine aviators pursuing their wings of gold. These students went on to fly in conflicts around the globe, including Vietnam and the Middle East. In 1991, the USS *Forrestal* (CV-59) was selected to replace the aging *Lexington* as the designated training carrier based at NAS Pensacola. This plan was subsequently revisited. In large part due to the expense of operation, pilot carrier qualifications are no longer conducted on a dedicated training aircraft carrier. Today, it is more cost effective for active fleet carriers to train student naval aviators while at sea off U.S. coastlines.

When decommissioned in 1991, the USS *Lexington* was the oldest operating aircraft carrier in the U.S. Navy fleet, leaving a number of legacies. Among them, the *Lexington* was the last *Essex*-class aircraft carrier on active duty and the first to have female crew members permanently stationed on-board. Today, the *Lexington* is permanently anchored in Corpus Christi, Texas. It serves our nation as a museum honoring all naval aviation personnel. The flight deck and hangar house a collection of aircraft including models that operated on the ship over its forty-eight years of service. The *Lexington* is one of five aircraft carriers currently preserved as public museums and the only one located on the Gulf Coast.

The P-8 Poseidon anti-submarine and maritime patrol aircraft replaced the P-3 Orion that served in this capacity since the 1960s. Squadrons operate from several air stations in the Gulf South, including NAS Jacksonville. The P-8's design is based on the efficient airframe of the Boeing 737-800 airliner. (U.S. Navy photo)

CHAPTER FIVE

TWENTY-FIRST-CENTURY CHALLENGES AND RESPONSES

The arrival of the new millennium brought new aggressors. Enemies once easily identified by their bold stands behind national flags were more often associated with disconnected political and cultural rebellions defying neat geographical boundaries. In response, military aviation evolved, developing tactics that had minimal impact on innocent civilians while defending democracy against threats spanning the globe. In the post–Cold War era as in years past, success in conflicts reflects the professional military training, facilities, equipment, infrastructure, and personnel utilized at a number of bases located in the Gulf South.

Not surprisingly, the U.S. military continues to favor this region as a center of operations for training in a turbulent geopolitical world climate. The Gulf South provides flexible training environments and is suited to staging scenarios in preparation for deployment of troops and equipment to remote areas. A number of bases along the Gulf Coast, often smaller and less well known than larger facilities, host specialized operations critical to the military mission. These bases and their personnel are creating a new contemporary legacy.

SPECIAL AIR OPERATIONS

The term "Special Operations" brings to mind revered military units such as the Navy SEALs, Army Rangers, or Delta Force, which are trained to respond to unique threats against our nation and its allies. The breadth and range of missions these personnel execute are vast, all of them depending on specialized military aviation operations, ranging from helicopters to transport aircraft for access to target areas. These flight crews often train on and are supported by bases in the Gulf South.

One such place is Hurlburt Field, Florida, home to the 1st Special Operations Wing (SOW), a unit of the U.S. Air Force. The unit itself is organized to execute specific operations, including maintenance, medical, and mission support. The wing is under the Air Force Special Operations Command (AFSOC), which, according to the Hurlburt Field website, provides "airpower to conduct special operations missions worldwide. The primary mission of the 1st SOW is to rapidly plan and execute specialized and contingency operations in support of national priorities. The wing's core missions include close air support,

precision aerospace firepower, specialized aerospace mobility, intelligence, surveillance and reconnaissance (ISR) operations, and agile combat support."

The Army features similar units. Fort Hood, Texas, Fort Polk, Louisiana, and Fort Benning, Georgia are three of the largest Army bases in the country that host special training and operations, each providing vast acreage and infrastructure to support practice for special missions as the needs arise. For example, Fort Polk operates the Joint Readiness Training Center Operations Group, or JRTC. With aviation support provided by the 5th Aviation Battalion, the mission of the center is "to provide installation support for power projection, combat readiness and mission execution for all tenant units as well as JRTC rotational units; and provide quality services and facilities." Fort Polk hosts continuous, year-round training exercises and provides units from other forts and bases with the needed resources for mission-specific immediate responses to terrorism and global threats to U.S. security.

2014. The UH-1N Huey (*right*) escorts a UH-1Y Venom (*left*) during the final flight of the UH-1N Huey for Marine Light Attack Helicopter Squadron 773, Marine Aircraft Group 49, at Naval Air Station Joint Reserve Base, New Orleans. The escort symbolizes the stepping aside of one helicopter platform to allow an improved version to take its place. (Photo courtesy of the U.S. Marine Corps/Cpl. J. Gage Karwick)

Opposite: A KC-135R Stratotanker assigned to the 6th Air Refueling Wing, 91st Air Refueling Squadron, at MacDill Air Force Base flies over central Florida. The KC-135 also provides refueling support to the U.S. Navy, U.S. Marine Corps, and allied aircraft. (U.S. Air Force photo/ Master Sgt. Keith Reed)

U.S. Army Blackhawk helicopters operate from numerous forts in the Gulf states, serving a variety of special missions. (U.S. Army photo/Sgt. Taresha D. Neal)

Staff Sgt. Adam Davila, 2nd Squadron, 38th U.S. Cavalry Regiment, exits a CH47 Chinook helicopter during a free fall jump at Fort Hood, Texas. (U.S. Army photo/Sgt. 1st Class Nicholas Ford)

2012. MH-53 Pave Low helicopters from Hurlburt Field, Florida, fly over the Gulf of Mexico. (U.S. Air Force photo/Senior Airman Emily Moore)

Special operations training at Hurlburt AFB. (Photo courtesy of the U.S. Air Force)

SHARED AIRCRAFT TECHNOLOGY

In the twentieth century, a number of aircraft types served all military branches, but this was due more to availability than to a calculated effort to have similar models operate across the services. For instance, at the start of World War II, the Stearman N2S biplane was available in abundance following the company's acquisition by Boeing. The Model 75 trainer was chosen by both the Navy and Air Force for primary training. In the pioneering days of aviation, though, the Army and Navy had their own ideas about the type of aircraft most beneficial to their respective operations. As a result, the branches developed more of their aviation programs separately rather than in unison.

Following World War II and through the Cold War, attempts at commonality met with mixed success, creating as much tension between the services as productive outcomes. The Navy continued its use of Grumman "Cat" family fighters, while the Air Force invested in the Century series of fighters—the F-100 through the F-106, produced by multiple manufacturers.

Today, there is a far more cooperative approach to stocking the U.S. military aircraft inventory. Cost is a major contributor to the shrinking variety of aircraft models in the modern military arsenal. Now more than ever, aircraft design specifications requested by the Department of Defense begin with the intent to cross over the services to the greatest extent possible rather than operate within a single branch.

In the twenty-first century, one of the most visible examples of inter-service aviation cooperation is the development of the F-35 fighter program. This aircraft and its capabilities are being incorporated into every branch with fighter aircraft inventory, allowing commonality in flight and ground crew training. Although each F-35 aircraft is constructed on a universal airframe, individual service models are delivered, such as the F-35A Air Force fighter model, the F-35B Marine Corps vertical takeoff and landing fighter, and the F-35C carrier-ready Navy version. Maintenance of the aircraft remains universal, with maximized interchangeability between the services, allowing cost reduction in parts and other resources. Eglin Air Force Base, near Shalimar and Valparaiso, Florida, is the center of operational development for the F-35 and is home to the aircraft's Integrated Training Center. The Navy and Marine Corps have personnel stationed at Eglin to expand this inter-service cooperative. From development to active service, each model and modification of this fighter of the future undergoes operational testing at Eglin.

A Navy F-35C, a Marine Corps F-35B, and an Air Force F-35A participate in a training sortie together at the 33rd Fighter Wing F-35 Integrated Training Center, near Eglin AFB. (U.S. Air Force photo/Staff Sgt. Katerina Slivinske)

Top: Dobbins AFB maintenance airmen work on an F-35 Joint Strike Flyer that landed at the Georgia base. The F-35 is built by the Lockheed Martin plant in Fort Worth, Texas. (U.S. Air Force photo/Don Peek)

Bottom: International cooperation: Royal Air Force pilot Hugh Nichols lifts off in a Marine F-35B Lightning II for the first "short take-off and vertical landing" (STOVL) sortie by a United Kingdom pilot at Eglin AFB. The U.S. Marines began flying STOVL sorties at Eglin in October 2013. (U.S. Air Force photo/Samuel King Jr.)

2014. Capt. Orlando R. Samudio (*left*) and Capt. Mike A. Jordan, both naval aviators with Marine Aerial Refueler Transport Squadron 252, pilot a KC-130J Super Hercules during tactical navigation training above Hunter Army Airfield, Georgia. The training allows pilots to hone skills of flying near demanding terrain and avoiding ground threats. (Photo courtesy of the U.S. Marine Corps/Cpl. J. R. Heins)

The C-130J Hercules is flown by every branch of the U.S. armed services. Aircraft assigned to the Air Force Reserve Command's 403rd Wing 53rd Weather Reconnaissance Squadron—the "Hurricane Hunters"—call Keesler AFB, Mississippi, home. These "citizen airmen" fly many hours not only over the Gulf of Mexico but also across the globe. Saltwater and moisture that could lead to corrosion are removed by an innovative washing system on the ramp at Keesler. (U.S. Air Force photo/Tech Sgt. Jame Pritchett)

TOP GUN—GULF COAST STYLE

Unknown to many residents of the Gulf South, the skies over Florida's emerald waters are one of the nation's most active, intensive training arenas for U.S. military fighter pilots. Miles from the shoreline, out of the way of civilian and commercial aircraft, the world's most advanced fighter and attack jets fly maneuvers simulating the heat of aerial battle. Participating aircraft fly to and meet within restricted airspace. Each pilot engages in a number of movements replicating actual conditions and combat that will be experienced in hostile regions around the globe. The exercises are critiqued by experienced flight instructors who have a breadth of knowledge in the evolution of aerial combat and who are specialists on the capabilities of each fighter aircraft that the students will encounter while deployed in combat areas. The end result is a highly trained aviator prepared for action in these conditions.

Combat between aircraft in flight has been known as dogfighting since the days of World War I. Today, air-to-air combat between fighter pilots has been refined to a scientific mix of aptitude and specialized education. It is a complex operation that is no longer simply a matter of "seat of the pants flying," or of acquiring skill through experience in meeting the enemy in the air. It is a combination requiring no less grit than that possessed by combat pilots of the past, but with the added element of mastery over a number of sophisticated technologies on board advanced aircraft, unheard of when the pioneers of flight first took to the skies.

Naval Air Station Key West, Florida, has become a centerpoint of advanced fighter and attack education for U.S. and allied pilots. The base is host to a range of training, from classroom instruction to complex computer systems that track every mile and maneuver of the pilots flying these demanding combat scenarios over the open water of the Gulf of Mexico. The training is similar to the famous Top Gun exercises brought to public attention by the movie of the same name.

A short-lived school of thought among a number of military contractors held that the rapid development of jet aircraft in speed and performance would make dogfighting a romantic image of the past. The idea became so prevalent that by the mid-1960s, many of the fighters

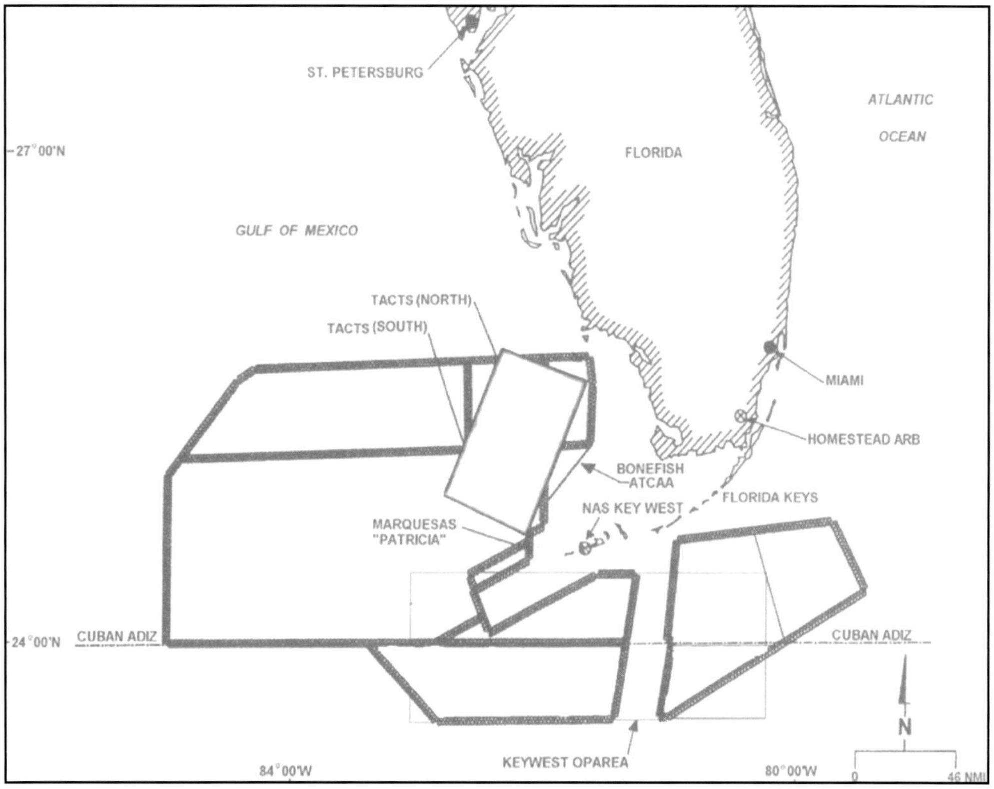

A grid map outlining the combat and practice flight areas surrounding south Florida. Aircraft exercises are conducted in offshore Military Operations Areas (MOAs), which are restricted or off limits to commercial and general aviation civil aircraft. (U.S. Navy illustration)

Opposite: **2014.** A plane captain places his hands on his head to advise the pilot that maintenance personnel are below the aircraft. This VFC-12 F/A-18C supports the NAS Key West Strike Fighter Advanced Readiness Program, a state-of-the-art facility for air-to-air combat training. (U.S. Navy photo/Mass Comm. Specialist 1st Class Brian Morales)

delivered to squadrons, including the F-4 Phantom II, were not originally equipped with guns. This mindset gained strength to the disgust of many fighter pilots, who argued persistently that the forecast was far from accurate.

One of the reasons behind the belief was the advancement of electronic warfare. Fighters were now assisted by other radar-carrying aircraft that had a specific mission of identifying enemy aircraft and other targets before they were in visual range. Another reason was the advent of heat-seeking missiles and smart weapons fired on enemy aircraft.

Although advanced airborne radar operations—delivered from a variety of aircraft, such as the Lockheed EC-121 Constellation (operated by both the Navy and Air Force)—could detect and identify enemy aircraft far outside of a pilot's visual range from the cockpit, interception by the pilot was not eliminated. By the end of the 1960s, the Defense Department readily accepted that the dogfight was not becoming a victim of tactical evolution. Large-caliber machine guns were installed on the F-4 fleet and other fighters lacking them. New aircraft orders included guns as a standard provision.

NAS Key West shares its education mission with fighter pilot schools in other parts of the nation. The U.S. Air Force created a similar program, known as "Red Flag," which became an advanced fighter training program based at Nellis Air Force Base, Nevada. However, NAS Key West uniquely specializes in realistic open-water air combat training. This concept was first used in the 1960s at NAS Miramar, California, near San Diego, as a means of transferring knowledge of experienced air combat veterans to pilots who would soon find themselves in the throes of a dogfight. The Navy began the program using the F-8 Crusader and the F-4 Phantom II fighters, the principal fighter aircraft in the fleet at that time.

Key West's program is comprehensive, incorporating specialized training for squadrons from bases across the United States. It also conducts combat exercises for National Guard pilots, as well as other service branches outside of the Navy and Marines. In addition, there are numerous allied and NATO units that regularly participate in these exercises in Florida's Gulf skies. The end result is that NAS Key West provides critical experience for all nations defending freedom and responding to air threats throughout the world.

THE B-52S OF BARKSDALE AIR FORCE BASE: STILL AMERICA'S BOMBER

Once the backbone of delivery for the Strategic Air Command, the Boeing B-52 bomber was stationed at numerous bases in the United States and allied nations across the globe. Today, only two domestic bases are home to these mighty aircraft: Minot Air Force Base in North Dakota and Barksdale Air Force Base near Shreveport–Bossier City, Louisiana.

Designed following the founding of the U.S. Air Force itself, the B-52 is the oldest active jet aircraft in the Air Force inventory. Flight of the first experimental aircraft, XB-52 and YB-52, took place in 1951. The first B-52A delivered to the Air Force rolled off the Boeing assembly line in April 1952 in Seattle, Washington, with the last aircraft manufactured in October 1962 at the Boeing facility in Wichita, Kansas. The B-52 Stratofortress has been upgraded and modified almost continuously since the last production H-model was delivered, making it the oldest among the few aircraft in continuous military service for over fifty years.

It is remarkable to consider that at the heart of a military operation, when nearly all the odds for victory depend upon the most modern of technologies, an aircraft surpassing well over a half century of service remains on the front line of democracy's arsenal. Records indicate that three generations of family military pilots—grandfather, father, and son—have logged time in the exact same aircraft over the years. The same is true of line and maintenance personnel who have kept the aircraft flying through the decades at bases throughout the United States.

In spite of new aircraft coming on the scene and ongoing technical innovation, the B-52 has been the U.S. Air Force's primary strategic bomber, and all indications are that it will remain so through the year 2040 or longer. It was flown during the time of the Cuban Missile Crisis in October 1961, during the Vietnam War in the 1960s and early 1970s, during the conflicts in Kuwait, Iraq, and other areas of the Persian Gulf in the 1990s, and during the war in Afghanistan in the twenty-first century. In January 1991, B-52s from Barksdale AFB flew nonstop round-trip, with aerial refueling, to bomb targets in Iraq on a mission lasting over 35 hours. The crews traveled 14,000 miles before returning to their home base in Louisiana. Such missions have been repeated in the course of the recent Middle Eastern conflicts.

As of 2014, approximately 70 of the 744 B-52 aircraft manufactured through 1962 were still flying for the Air Force. The majority of the B-52 fleet was retired to Davis-Monthan Air Force Base, known as the nation's aircraft "Boneyard," in Tucson, Arizona. Beginning in 1990, over 300 B-52s were destroyed in the Boneyard in full view of Soviet satellite cameras in accordance with SALT 1 (Strategic Arms Limitations Talks). The remaining aircraft were lost through the years in combat or operational accidents.

Providing support for ongoing conflicts around the world, Barksdale AFB continues to be the center of southern operations for the remaining B-52 fleet.

2014. B-52H Stratofortress aircraft sit on the flight line at Barksdale AFB. B-52 bombers are based at Barksdale and at Minot AFB in North Dakota. (U.S. Air Force photo/Airman 1st Class Benjamin Raughton)

Opposite: A B-52 Stratofortress, flown by Capt. Will Byers and Maj. Tom Aranda, prepares for refueling over Afghanistan during a close-air-support mission. The crew is deployed from the 2nd Bomb Wing, Barksdale AFB. (U.S. Air Force photo/Master Sgt. Lance Cheung)

EGLIN AIR FORCE BASE: FROM GRAVITY BOMBS TO SMART BOMBS AND ADVANCED AIR-TO-AIR MISSILE TECHNOLOGIES

Land warfare and air combat evolved in unison through the first one hundred years of military flying. As the design and utility of aircraft have improved, so have the targets sought by these aircraft: underground pathways, storage, and living quarters, all designed to secure assets. Like the tunnels of the American Revolution and nineteenth-century European battlefields, the trenches of the First World War were little more than an extensive series of open paths to connect redoubts along an infantry battle line. World War II introduced stronger facilities to prevent a repeat of trench-combat slaughter. Bunkers both on the battlefield and behind the lines were designed to protect personnel from attacks. From the Korean and Vietnam conflicts through those in the first decades of the twenty-first century, especially against Iraq, Afghanistan, and independent terrorist organizations, challenges continue to be encountered.

Throughout its existence, including operations as the Valparaiso Gunnery and Testing Range, Eglin Air Force Base, Florida, has been the principal and center for incorporation of air-launched weaponry. Today, flight testing and development of air-to-air and air-to-ground unguided rockets, guided missiles, and other precision-guided munitions are continuous at Eglin, thus ever expanding the nation's defense capacity to locate and penetrate aggressive barriers and fortifications, and to disable the aircraft and vehicles supporting their operation.

There exist numerous versions of what the public and media generically refer to as "bunker bombs" and "smart bombs." Each is individually designed for specific targets hidden anywhere, from within populated areas to isolated remote desert outposts. These weapons can be delivered by aircraft from all military branches, from long-range Air Force B-1 and B-52 bombers to Air Force fighter and Navy carrier attack squadrons.

Gen. Henry "Hap" Arnold, chief of the Army Air Forces, incorporated precision-guided munitions into combat operations during World War II. According to Dr. Robert Kane, Air Force historian at Maxwell Air Force Base:

> By late 1945, the program had produced very few successful weapons because of the state of technology (vacuum tubes verses microchips and transistors). The program did develop the Vertical Bomb (VB) 1 AZON (azimuth only), which successfully attacked rail lines and tunnels in Nazi-occupied Holland and in Japanese-held territories in Burma in the latter months of World War II. In summer 1944, Arnold also had captured German V-1 "Buzz Bombs," which were shipped to Eglin for testing and resulted in development of the Jet Bomb (JB) 1, called the "Loon," predecessor to today's cruise missile. The VB-4 RAZON (Range and Azimuth only), was used during the Korean War.

Today, Eglin leads the military in the ongoing development and operational testing of the electro-optical Maverick missile, laser-guided technologies of the Pathway series, and various GPS-guided weapons such as the Joint Direct Attack Munition (JDAM) and the

Small Diameter Bomb (SDB) and their variants. The MOP, or Massive Ordnance Penetrator, is one of the most effective responses to contemporary threats from underground bunkers and hidden resources across the globe. The range of these weapons allows flight crews to remain outside of targeted areas, and following launch, they require little or no assistance in completing their mission.

In addition to weapons testing at Eglin, the Gulf Coast also hosts production of the Joint Air-to-Surface Standoff Missile–Extended Range (JASSM–ER) weapons, among others, at the Pike County Operations' Long Range Strike Systems annex, near Troy, Alabama.

An explosion left from a Joint Direct Attack Munition propels debris in the air during a test. The 780th Test Squadron at Eglin AFB conducts air-to-air, air-to-ground, and ground testing on munitions. (Air Force photo)

Opposite: **2009.** A B-52 bomber, escorted by an F-16 chase aircraft, releases a test version of the Massive Ordnance Penetrator over White Sands Missile Range, New Mexico. (Department of Defense photo)

In the early 1960s, U.S. Navy and Marine pilots trained in Grumman F9F Cougars. (U.S. Navy photo)

CHAPTER SIX

CONTEMPORARY FLIGHT TRAINING

From the original Wright Brothers' aircraft, Curtiss Pushers, floatplanes, and Jennys, military aviation in the course of one hundred-plus years has evolved to technologies unimagined by those pioneering pilots who first took to the skies. Throughout that entire span of time, the Gulf Coast has been a home for flight training conducted by every branch of the U.S. military, hosting courses ranging from primary and advanced instruction to squadron- and mission-specific training at a number of bases in the region. The reason is simple. The Gulf South is an ideal location for student pilots, offering—among a number of other benefits—a favorable climate lacking in harsh winters, an exceptional mix of terrains, and proximity to open water. It has proven itself over the past century a perfect environment for the demands of the task.

After primary and advanced instruction, military pilots and flight officers move on to other regions of the country to specialize in operating more complex aircraft. They then train in any number of climate conditions encountered across the globe, from biting Arctic cold to extreme dry desert hazards, situations experienced in active deployment. However, the critical introduction to a career in military aviation begins in the Gulf South.

EARLY EVOLUTION AND ECONOMICS

Although the principles of flight have varied little from military branch to branch, when it comes to actual pilot instruction, until recently, there was minimal interaction between the service training programs. Each branch operated separately with an independent syllabus. Justification for this segregation was as much a matter of competitive funding as it was of service mission. Each aviation program received its own dedicated budget, a resource not so cordially shared with its counterparts.

At the same time, there existed a commonality of equipment and aircraft used in primary pilot training. Just prior to the outbreak of World War II, the Link flight training simulator was produced in great numbers and incorporated into the curriculum of both Army and Navy primary and instrument instruction. Another shared resource was the Stearman biplane, affectionately remembered by its student pilots as the "Yellow Peril." It evolved from the older biplane designs and was flown by both the Army Air Corps and the Navy as a primary training aircraft. In 1943, on the runways of NAS Corpus Christi, future president George H. W. Bush, piloting a Stearman, became the youngest naval

Primary training helicopters such as this Bell model introduced countless U.S. Army warrant officers to military aviation in the skies of central and south Alabama. (U.S. Army photo)

aviator to receive wings. In the postwar aviation collector's market, airworthy Stearman aircraft have survived in abundance, becoming a favored icon representing primary training of naval aviation cadets and Army Air Corps student pilots throughout World War II.

By 1941, other aircraft manufacturers pursued the now-booming military training aircraft market. Consolidated-Vultee Aircraft Company and North American Aircraft Company, among others, began producing a new generation of monoplane (single-wing) training aircraft, the SNV and SNJ models, respectively, to replace the aging biplane design. These trainers were delivered in large numbers to the Gulf Coast bases that were rapidly constructed during mobilization for World War II. Once pilots graduated from primary training, assignments relocated them throughout the country, with further introduction to multi-engine and high-performance aircraft in preparation for deployment to the European or Pacific theater.

Progress in training technology and new aircraft that had begun during the war continued into the next decade. In the 1950s, a rapid transition to the jet engine got underway, though a number of combat squadrons continued to operate highly advanced piston-engine aircraft brought online after the war ended, arriving too late for active squadrons to utilize.

Increasingly complex and powerful aircraft forced the expansion of airfields. NAS Pensacola quickly outgrew the utility of Chevalier Field. In 1950, this historic property was still operating within the original acreage and design constructed for Station Field three decades earlier. By 1955, landlocked and its runways unable to be extended for use by jet technology, Chevalier was replaced by the much larger Sherman Field, located to its west and on the opposite side of the Florida base. Similarly, Army Air Force bases across the Gulf South were in dire need of expansion to support new fighter and bomber designs with heavier takeoff weights requiring longer runways. Bases extended and widened their aircraft-operating areas, increasing runway lengths up to 10,000 feet to amply accommodate jets and larger aircraft.

Chase Field in Beeville, Texas (closed in 1993), provided advanced flight instruction for Navy, Marine, and Coast Guard aviators. The property is now Chase Field Industrial Park, a municipal property development. (Photo courtesy of the U.S. Navy)

An Army student pilot prepares for a training flight in a Blackhawk helicopter at Fort Rucker. (U.S. Army photo)

2012. At NAS Corpus Christi, Marine Corps Capt. Chris Latimer (*left*), an instructor pilot for Training Squadron VT-31, and Ensign Marvin Smith, a student naval aviator, conduct pre-operation procedures before a training flight in a T-44C Pegasus aircraft. (U.S. Navy photo/ Richard Stewart)

Capt. Damien Ludwick, 559th Flying Training Squadron, completes a pre-flight check prior to a training flight in a T-6 Texan II at Randolph AFB. (U.S. Air Force photo/Dave Terry)

Ensign Christopher Tucker, a student naval aviator assigned to Training Squadron VT-28 at NAS Corpus Christi in 2011, fastens his SV-2 survival vest before his training flight in a T-34 Turbomentor aircraft as instructor pilot Lt. Cmdr. Daniel Kelly looks on. (U.S. Navy photo/ Richard Stewart)

One of many examples of the efficiency produced by cross-utilization of military branch air base facilities is the U.S. Air Force 479th Flight Training Group, now permanently based at NAS Pensacola. (Photo courtesy of the U.S. Air Force)

A T-6 Texan II prepares for landing at Randolph AFB. (U.S. Air Force photo/Don Lindsey)

A pair of T-6 Texans practices formation flying above the Gulf Coast. (Photo courtesy of the U.S. Air Force)

CONTEMPORARY TRAINING

The latter half of the twentieth century witnessed increased ground-school and academic classroom time for all U.S. armed services' aviation training. The additional coursework reflected the complexity of post–World War II subject matter and the changing environment of human flight, particularly the introduction of jet aircraft. A typical contemporary air training operation for all the services includes a main training base surrounded by smaller training airfields within a 50-to-100-mile radius. The remote airfields allow primary training students to learn the principles of flight, takeoff, landing, and basic navigation on short distances.

Navy

NAS Pensacola is still known by the title it earned many years ago, the "Cradle of Naval Aviation," a proud legacy that generations of naval aviators never outgrow. In years past, the initial naval aviator (including pilot and flight officer) candidate school was conducted at this historic Gulf Coast naval air station. Today, however, naval aviation officer candidates, as well as all non-aviation naval officers, begin their training with the Navy's Officer Training Command program at Naval Station Newport, Rhode Island. Following this initial step, officers pursuing aviation are assigned to one of the Gulf Coast air stations to begin flight training. The contemporary naval aviation flight training mission is shared among NAS Pensacola and NAS Whiting, Florida, NAS Meridian, Mississippi, and NAS Corpus Christi and NAS Kingsville, Texas. Other naval bases, while not funded specifically as training bases, support the operations of these primary airfields.

The U.S. Navy educates its personnel through the Naval Education and Training Command, which is headquartered at NAS Corpus Christi and has a presence at all Navy facilities. Serving as a university system within the service, the Command coordinates all personnel education in and outside of aviation. Under the Command's umbrella, the Navy operates the Center for Naval Aviation Technical Training, with attention also given to enlisted and officer personnel that are not pilots. The Naval Air Technical Training Center, specifically for maintenance personnel, is at NAS Pensacola and is one of the largest facilities within the Training Command. NAS Pensacola is also home to ninety Department of Defense operations facilities.

The Command's Gulf Coast facilities also train Marine aviation personnel. Marine pilots train with naval aviation students in Florida and then transition to mission-specific Marine bases elsewhere within the Corps. Prior to closing the field and constructing the training center on the site, NAS Pensacola transformed the original Chevalier Field into the Navy and Marine Corps' premier helicopter pilot training facility.

Student pilots from the Navy, Marines, Coast Guard, and Air Force often share simulators and instructors. Commonality of the primary training aircraft and course syllabus has reduced cost and increased interaction between the services. (Photo courtesy of the U.S. Air Force)

2013. Tedd Muery (*back left*), a T-45 simulator instructor, and LTJG Hadley Sulpizio, a physiologist from the Naval Survival Training Institute (NASI), give dynamic hypoxia training to Lt. Eric Schwab, an instructor pilot in Training Squadron VT-21. The hypoxia training involves a reduced oxygen breathing device that exposes the air crew to actual hypoxic conditions. (U.S. Navy photo/Richard Stewart)

2002. At NAS Pensacola, Hospital Corpsman 2nd Class Tim Sudduth demonstrates the Virtual Reality (VR) parachute trainer as Aviation Survival Equipment Specialist1st Class Jackie Hilles controls the program below. Students wear VR glasses while in a parachute harness and learn to control their movements through computer-simulated scenarios. The VR trainer also teaches air-crew personnel how to handle a parachute in different weather conditions and during equipment malfunctions. (U.S. Navy photo by Chief Photographers' Mate Chris Desmond)

NAS Pensacola students at the Aviation Rescue Swimmer School undergo a parachute disentanglement exercise. The physically arduous four-week course trains enlisted air crewmen as rescue swimmers for search-and-rescue operations. (U.S. Navy photo/Chief Photographers' Mate Chris Desmond)

Hospital Corpsman 2nd Class Jerry Morrish demonstrates NASI emergency ejection-seat training, NAS Pensacola. (U.S. Navy photo/Chief Photographers' Mate Chris Desmond)

Today, the grounds of the original Station Field, renamed Chevalier Field, at NAS Pensacola are home to the Naval Air Technical Training Center, providing instruction to all branches of the U.S. military and to allied forces. (Photo by author, courtesy of the Public Affairs Office, NAS Pensacola)

Above: Col. Tom Murphy, 47th Flying Training Wing commander, and Lt. Col. Steve Plank, 434th Fighter Training Squadron, director of operations, prepare to exit the T-38C Talon they delivered from Laughlin AFB to the 435th Fighter Training Squadron at Randolph AFB. The aircraft is the last of twenty jets transferred in 2012 from the 434th as part of the consolidation of fighter fundamentals training. (Photo courtesy of the U.S. Air Force)

Opposite: **2012.** A naval aviation student and instructor following a training flight at NAS Meridian. (Photo courtesy of the Public Affairs Office, NAS Meridian)

Coast Guard

The U.S. Coast Guard operates the Aviation Training Center in Mobile, Alabama, under the jurisdiction of its 8th District, headquartered in New Orleans, Louisiana. The Coast Guard flies both fixed-wing patrol aircraft and rescue helicopters. The facility educates personnel on operations and technical matters relative to the HC-144 and HC-130 fixed-wing aircraft, MH-60 and MH-65 helicopters, unmanned aircraft systems, ship-to-helicopter operations, rescue swimmers, and special missions. This specific training relative to homeland security and other missions is coordinated through the Alabama facility.

Air Force

Since the days of Kelly Field and the earliest Army Air Corps training programs, the U.S. Air Force has maintained its historic association with San Antonio and other training facilities throughout Texas, including, of course, Randolph and Lackland, and also Sheppard Air Force Base, located in Wichita County. These facilities have a primary mission of training pilots and aviation specialties. Air Force pilots receive instruction at other facilities in the South as well. To the east, Columbus Air Force Base in north Mississippi is home to a large training wing providing primary flight instruction to budding Air Force pilots.

The Air Force operates the Air Education and Training Command from Randolph Air Force Base to execute a variety of specialized instruction. Another Air Force institution of learning is Air University. Headquartered at Maxwell Air Force Base, Alabama, Air University is the intellectual and leadership center for the service, providing academics for nearly all disciplines in the Air Force to both officers and enlisted personnel.

Army

In the 1950s, Fort Rucker near Montgomery, Alabama, became the centerpiece of Army aviation, training warrant officer pilots and helicopter crews after the expansion of military rotary wing flight. The U.S. Army Aviation Center of Excellence at Fort Rucker conducts all Army rotary wing training and advanced instruction. Further specific training is conducted at other forts and facilities.

COMBINED SERVICE PROGRAMS

Over the years, especially in peacetime when funding sources were increasingly scrutinized, the military branches came to value the similarity of their flight training programs. While the mission of each service remained unique, the instruction process and personnel responsibility became ripe for consolidation.

Today, basic aviation training is divided among the Air Force, Army, and Navy at various air stations, bases, and forts. Together, the branches have explored ways to maintain quality while sharing flight schools, simulators, aircraft, and other infrastructure, a cost-effective measure for taxpayers and for the military branches themselves. What has not changed is the place where tomorrow's military pilots and flight crews are first introduced to the mission of national defense: the states of the Gulf South. The Gulf Coast flight training facilities have largely survived the financial strain that resulted in the closure of their counterparts elsewhere in the country.

Cost-effective pilot and support-personnel cross-training has been initiated in a number of military career paths, especially for primary programs and advanced flight instruction. Specific training remains specialized for the particular branch served and the individual mission of the squadron personnel. One of the most visible signs of this cooperative approach is the use of the T-6 Texan II, manufactured by Raytheon (Beechcraft) Aircraft Corporation, for basic fixed-wing flight training by the Air Force, Navy, Coast Guard, and Marine Corps. Identical aircraft models and accompanying ground-school classroom instruction are utilized throughout the Gulf South on naval air stations and Air Force training bases.

MILITARY FLIGHT TRAINING: A RIGOROUS CAREER PATH

A career in military flying is both mentally and physically demanding. Success does not rely on political connections or heredity, but rather on hard work, talent, and sheer determination. The journey begins with officer candidate screening. The Air Force, Navy, Marines, and Coast Guard require pilots to be college graduates and commissioned officers. The Army's warrant officer program allows qualified personnel who have demonstrated aptitude but not completed college to apply for pilot training.

Applicants must possess the capacity for long-term commitment and academic challenges. The time from selection as an officer candidate to an active-duty squadron assignment is up to three years of ongoing training. However, education continues throughout a life in aviation. Commitment to service in an aviation field ranges from eight to ten years, depending on the branch. Every military pilot must demonstrate and maintain exceptional physical conditioning and be a better-than-average swimmer.

All service branches first indoctrinate selected candidates in the fundamentals of leadership. The reason is simple. Pilots are responsible for the lives of their crew and often exercise sole discretion in operating multi-million-dollar aircraft. Acceptance in the program does not guarantee success. In the course of instruction and for a variety of reasons, ranging from health issues to aptitude, numerous candidates (sometimes as many as half or more of the original class) do not complete the flight program.

Basic instruction and primary and intermediate flight training last approximately one year. In this stage, the student must master aviation sciences related to, among other disciplines, aerodynamics, meteorology, physics, and physiology. A passion is necessary for the physical demands experienced while flying outside of straight and level, point-to-point operations, including maneuvers exceeding and reversing the pressures of gravity, popularly referred to as "pulling Gs."

Each military branch has an advanced training program geared to

2012. Lt. Col. John Turnipseed (*left*), 559th Flying Training Squadron instructor pilot, and Air Force Academy foreign exchange student Cadet Arturas Kucharevas of Lithuania perform final checks before an orientation flight in a T-6 at Randolph AFB. (U.S. Air Force photo/Rich McFadden)

its own unique needs. The Air Force, Navy, and Marines all operate fighter and attack aircraft. All branches, including the Coast Guard, operate versions of the C-130 Hercules transport aircraft. The Air Force uniquely operates the nation's fleet of massive transport and cargo aircraft such as the C-5 Galaxy and the C-17 Globemaster. Army aviation specializes in the range of use and utility of the helicopter. Graduating flight students of each service, upon receiving their wings, request a preference but may be assigned a career path based on the needs of the branch. Mission-specific training may take an additional year before the pilot is assigned to a squadron.

The career path in naval aviation is a long and especially challenging one. Like Air Force pilots, student naval aviators spend hundreds of hours mastering basic and advanced flight skills in both turbine and jet aircraft. However, within the ranks, select naval aviators are trained to operate from the decks of aircraft carriers moving through the world's oceans. This group will uniquely "trap" their shipboard landings with a tailhook—the dramatic snagging of the vessel's arresting cable by a hook on the rear of the aircraft. Not all naval aviators are trained in carrier operations. For those in land-based operations, their mission, while perhaps lacking the popularized Hollywood public appeal, is no less critical to national defense.

INTERNATIONAL STUDENT TRAINING

The continental United States has been an attractive location for flight training of allied pilots since the 1930s. Prior to December 7, 1941, when the United States entered World War II, and during the conflict itself, American military bases were away from the hostilities of foreign battlegrounds and provided a vital isolation, allowing these nations to train their personnel in security.

A number of international pilot training programs begun during World War II continued throughout the 1950s and 1960s. The U.S. Air Force hosted programs for allied nations and NATO members in the Gulf South. During the Vietnam War, Sheppard Air Force Base, Texas, was a major jet training center for Laotian and Vietnamese air force officers who traveled to the United States to attend flight school.

Similarly today, a number of countries do not have comparable facilities of their own. U.S. allies in Europe, the Middle East, the Pacific region, the Americas, and on the African continent share U.S. facilities, conducting their primary and advanced training at bases in the United States. Extension of United States military resources to train allied and other friendly nations cooperating in defense of global freedom is a potent symbol and means of democracy at work.

Right: Students from allied countries travel to Gulf Coast training bases for specific instruction. Here a student from the French Fleet Air Arm (Aéronavale) attends courses at NAS Meridian. (Photo courtesy of the Public Affairs Office, NAS Meridian)

Below: **2006.** Capt. Alberto Surace (*left*) and 2nd Lt. Mathew Spinelli (*right*), students of the Euro-NATO Joint Jet Pilot Training program, prepare for takeoff in a T-38 Talon at Sheppard AFB, Texas. (U.S. Air Force photo/Tech. Sgt. Cecilio M. Ricardo Jr.)

Left: **2010.** From ground level, the "Taj Mahal" tower reminds student pilots and base visitors of the heritage of Randolph Field. (U.S. Air Force photo)

Below: **2010.** This restored hangar on the original site of Brooks Field, Texas, is one of the treasured remaining buildings from World War I in the San Antonio area. (U.S. Air Force photo)

JOINT BASE SAN ANTONIO: LIVING HERITAGE

Through a series of base closures and realignments in 2005, Randolph Air Force Base and other historic bases in the San Antonio area, including the resources of Brooks as well as other remnants of the Kelly network from 1915 through the 1920s, became united in a "joint base" now referred to as Joint Base San Antonio, or JBSA. The joint base operation includes the contemporary facilities of Fort Sam Houston, Lackland AFB, and Randolph AFB, as well as outlying facilities supporting these base operations. The idea was to have a single administration and command that would ensure minimal duplication by the bases and by the branches of service occupying them. This single command is under the administration of the U.S. Air Force.

While NAS Pensacola's historic structures face threat from hurricanes, the risk to JBSA's heritage comes from operational relocation and abandonment, a pressing need for real estate, and contemporary updates to the facilities. The Air Force has done an outstanding job of preserving its heritage. Through the years, a number of historic buildings on Randolph as well as other bases in the San Antonio area have been placed under permanent protection. The Air Force continues to keep the majority of these surviving facilities active and vibrant. Randolph is now the proud operator of a historic district created to feature these irreplaceable buildings on the fields that have become an important part of Air Force culture, inspiring new members of the armed services to appreciate the pioneers that paved their path.

From the "Taj Mahal," the chapel, the main avenue and gate, and the original hangars, to the simplest surviving original base housing and quartermaster offices, as well as other historic buildings constructed in the days before World War II, these ever-present reminders of the birth of the modern Air Force remain to be seen by student pilots, flight crews, and the far more numerous ranks of support and technical personnel who receive their training at JBSA facilities.

It is critical that future generations understand the role these buildings played in the preservation of freedom across the globe during World War II, Korea, Vietnam, smaller regional conflicts, and Middle East actions, as well as in support of the ongoing development of our contemporary Air Force. These buildings also memorialize the significant role that the San Antonio area played in training military personnel for numerous allies since the birth of flight.

BLUE ANGELS CALL THE GULF COAST HOME

Since their founding in 1946, the U.S. Navy Blue Angels have been based exclusively at Gulf Coast naval air stations, including Jacksonville, Florida, Corpus Christi, Texas, Whiting Field, Florida, and Pensacola, Florida. The U.S. Navy Flight Demonstration Team, designated an official active Navy squadron, is by any measure the most famous and popular air show attraction in the United States, if not the world. The Blue Angels are featured in approximately forty or more air shows around the globe each year between March and November, performing before millions of fans and serving as one of the Navy's principal recruiting resources.

The team was founded a year after the conclusion of World War II by naval aviators under the authority of Adm. Chester Nimitz, chief of naval operations and victorious leader of the Pacific theater. Nimitz supported the team's desire to demonstrate combat piloting to citizens on the home front who had not had an opportunity to witness naval aviation in action. With the exception of the Korean War years and a federal sequestration in the 2013 season, the Blue Angels have performed annually since their inception.

In every contemporary show, over thirty specific maneuvers are performed by six Angels aircraft: four jets in diamond formation and two jets accompanying in solo demonstration. Since 1970, the team has also included "Fat Albert," a U.S. Marine Lockheed C-130 Hercules support aircraft that transports operational staff and maintenance equipment necessary for each show. "Fat Albert" has been incorporated into the performances and become a popular attraction in its own right, demonstrating how these agile though large aircraft are capable of delivering supplies to front-line combat operations on short, unpaved runways.

In its many years of performing, the team has used a number of aircraft, including the F6F Hellcat, the F8F Bearcat, F-9 Panther, F-9 Cougar, F-11 Tiger, F-4 Phantom, A-4 Skyhawk, and the F/A-18 Hornet. Support aircraft operated prior to the C-130 include the Douglas R4D Sky Train, the Curtiss R5C Commando, the Douglas R5D Skymaster, and the Lockheed C-121 Super Constellation.

Preseason training and rehearsal take place outside of the Gulf Coast at Naval Air Facility (NAF) El Centro, California, considered the winter home of the Blue Angels. That facility offers a vast open area over unpopulated deserts for training new team members and reworking routines by experienced crews.

Included within a full national and international demonstration season are several shows at military bases throughout the Gulf Coast states. In addition, the Blue Angels feature two shows for their hometown audience. Each July, a performance on Pensacola Beach thrills the panhandle community. The team's final homecoming show of the season is over Sherman Field at NAS Pensacola in November, where naval aviators and support personnel are formally reunited with their families, concluding a year of dedication and difficult traveling.

The first female naval aviators, Marine Captains Katie Higgins and Corrie Mays, joined the Blue Angels squadron for the 2015 season, commanding the C-130 "Fat Albert."

1946. The original Navy Blue Angels Squadron in front of one of their F6F Hellcat fighters at NAS Jacksonville. (U.S. Navy photo)

December 6, 2012. Lt. Mark Tedrow, opposing solo pilot for the Blue Angels, performs a section high-alpha maneuver during a training flight over the Gulf Coast. (U.S. Navy photo/Chief Mass Comm. Specialist Russ Tafuri)

High above the Florida coastline, an F-35A Lightning II pilot navigates the aircraft toward an Air Force Reserve KC-135 Stratotanker from the 336th Air Refueling Squadron, March AFB, California. (U.S. Air Force photo/Master Sgt. John Nimmo)

AFTERWORD

Ask any U.S. military flyer, past or present, and you will find they have vibrant memories of training days on the Gulf Coast. For the past one hundred years, and now into the twenty-first century, along the runways and within the historic buildings of this region, they mastered the initial skills to become critical leaders in the most highly trained and prepared military force in the world.

With the infrastructure firmly planted in these regional bases and stations, student pilots and aviators of the foreseeable future will cherish challenging and critical periods of their development in the same way. Some, following their initial training, will continue their residency, assigned to squadrons based in Gulf states, where they will return following deployment overseas. Others will move on to bases elsewhere in the country for specialized advanced training. All retain and employ military aviation knowledge and experience gleaned on the Gulf Coast, stretching from the plains of Texas to the emerald waters of Florida.

Military aviation is an integrated, integral part of the culture of the Gulf states. Not a single day goes by without a Gulf Coast resident gazing upward and seeing a military flight crew at work in the sky on a mission originating from one of these bases. The economic impact of this presence on the region is in the billions of dollars annually. The historical evolution of these six states—Texas, Louisiana, Mississippi, Alabama, Georgia, and Florida—has in no small measure been influenced by the presence and residency, be it temporary for training or permanent assignment order, of thousands of aviation personnel stationed within their communities. Eternal vigilance being the price of liberty, the Gulf South states provide.

APPENDIX 1

MILITARY AIRFIELDS AND AIR STATIONS IN THE GULF SOUTH

The following is the author's attempt to provide the reader with a general, summarizing list of military airfields, including U.S. Army Air Corps, Army Air Force, Air Force, Navy, Marine, and Coast Guard aviation facilities, that have operated in the Gulf South states. I have yet to find a single listing of all such facilities, inclusive of all branches, that can truly be called comprehensive; this statement includes my own attempt presented here. Any omissions in this attempt are unintentional and entirely my own fault.

When I first set out to include this particular appendix, I thought it would be a simple, straightforward list. I was wrong. I was soon struck by the enormity of the task. The number of such operations over the past one hundred years is quite extensive, often associated with a complex service evolution involving changes in name, mission, branch lineage, decommissioning, and reactivation. Some facilities are referenced by name, while others are referenced by the metropolitan area in closest proximity.

Self-supporting air bases and air stations were accompanied by a multitude of neighboring dependent auxiliary facilities. Some but not all auxiliary facilities were formally named. Many were referred to simply by a designated number. Some housed no permanent personnel. Eglin Army Air Base, for example, operated ten auxiliary fields. Many were intended to be temporary, while others became independent, self-supporting bases and stations in their own right. Due to the large number of such facilities (approximately two hundred within the Gulf Coast states alone) and the challenge of appropriate identification, a complete listing of auxiliary fields is not attempted here.

A note on military property abbreviatons: Over the first one hundred years of military flight, numerous property abbreviations have been used to identify U.S. military air facilities. The following abbreviations have been used to identify these property sites, the majority of which through a transfer of command or function have used more than one designation in the course of operations. In addition, throughout this history, titles such as "Annex," "Center," "Field," and "Installation" have been incorporated (both officially and unofficially) into location names as a general identifier for aviation properties, or properties hosting aviation-related activity. Not all abbreviations applied to a particular military site are incorporated into this appendix.

ABBREVIATIONS

AAAF	Auxiliary Army Air Field
AAB	Army Air Base
AAF	Army Air Field
AB	Air Base
AF	Auxiliary Field
AFB	Air Force Base
AFS	Air Force Station
ANGB	Air National Guard Base
JARB	Joint Air Reserve Base
JRB	Joint Reserve Base
NAS	Naval Air Station
NAAS	Naval Auxiliary Air Station
USCG	United States Coast Guard Base

ALABAMA

Anniston AFB
*Aviation Training Center, USCG
Bainbridge Field
Barin Field NAS
Bates AAF
Birmingham ANGB
Brookley AFB
Cairns Field
Camp McClellan
Courtland AAF
Craig AFB
Decatur Field
Demopolis AAF
Fort Rucker
*Daleville Army Aviation Center, Fort Rucker
*Dothan Army Aviation Center, Fort Rucker
*Enterprise Army Aviation Center, Fort Rucker
*Ozark Army Aviation Center, Fort Rucker
Griel Field
Gunter AFB/AFS/Annex
Kennedy Field
Maxwell Field/AFB
Montgomery ANGB
Moton Field
Napier AAF
Reilly Field
Roberts Field
Shorter Field
Taylor Field
Tuskegee AAF
Van De Graaf AAF

FLORIDA

Alachua AAF
Avon Park Air Force Range
Baldstein AAF
Banana River NAS
Blancur AAF
Boca Raton AAF
Bronson Field NAS
Buckingham AAF
Carlstrom Field
Cecil Field NAS
Chapman Field
Corry Field NAS
Dale Mabry AAF
Daytona Beach NAS
Deland NAS
Dillon AAF (Choctaw)
Dorr Field
Drew AAF
Duke Field
Dunnellon AAF
Eglin AFB
Eighth District Headquarters, USCG
Ellyson Field NAS
Epler AAF
Fort Lauderdale NAS
Graham AFB
Hendricks AAF
Homestead ARB
Hurlburt Field
Jacksonville ANGB
Jacksonville NAS
Key West NAS
Lakeland AAF
Lodwick Field
MacDill AFB
McCoy AFB
Melbourne NAS
Miami AAF
Miami/Opa-locka NAS
Orlando AFB
Palm Beach AFB
Patrick AFB
Peel AAF
*Pensacola NAS
Piccolo AAF
Pierce AAF
Pinellas AAF
Richmond NAS
Riddle Field
Sanford NAS
Sarasota AAF
Saufley Field NAS
Tanner-Ewing AAF
Taylor Field
Tyndall AFB
Valentine Field
Venice AAF
Vero Beach NAS
Wagner AAF
*Whiting Field NAS

GEORGIA

Albany AFB/NAS
Athens Field
Atlanta AAF
Atlanta NAS
Babcock AAAF
Bainbridge AAF
Bemiss AAF
Berlin AAF
Bush Field
Camp Benning
Chatham AAF
Cochran AAF
Daniel AAF
Decatur AAAF
Dionne AF
Dobbins JARB
Donalsonville AF
Dorminey AAF
Douglas Field
Faceville AAAF
Glynco NAS
Green Cove Springs NAS
Gunn AAAF
Harris Neck AAF
Hayley AAF
Hunter Field
Knight AF
Lake Park AAF
Lawson Field
Leesburg AAF
Marietta AAF
Moody Field
Moultrie Airport AAF
New River AAF
Norman Park AAF
North Smithville AAF
Paulk AF
Perry AAAF
Reynolds AF
Reynoldville AAF
River AF
Robins Field
Rock Ford AAF
Souther Field
Spence Field
St. Simons NAS
Statesboro AAF
Thomasville AAF
Tifton Airport AAF
Turner Field AFB
Vada AAAF
Valdosta AAF
Vidalia-Lyons AAF
Waycross AAF
West Leesburg AAF
West Smithville AAF

LOUISIANA

Barksdale AFB
Baton Rouge AAF
Chennault AFB
DeRidder AAB
England AFB
Esler AAF
Fort Polk
Gerstner Field
Hammond AAF
Houma NAS
Lafayette AAF
Moisant AAF
New Iberia NAAS
New Orleans AAB
New Orleans NAS
Selman AAF
Seventh District Headquarters, USCG

MISSISSIPPI

Cleveland AAF
*Columbus AFB
Fletcher Field
Greenville AFB
Greenwood AAF
Grenada AAF
Gulfport/Pascagoula AAF
Hattiesburg AAF
Jackson AAF
Keesler AFB
Key Field AAF
Laurel AAF
*Meridian NAS
Payne Field

TEXAS

Amarillo AFB
Avenger Field
Barron Field
Bergstrom AFB
Biggs AFB
Brooks AAF
Brownsville AAF
Bryan AFB
Call Field
Camp Dick
Camp John Wise
Camp Mabry
Carswell AFB/NAS JRB Fort Worth
Chase Field, NAS
Childress AAF
*Corpus Christi NAS
Corsicana Field
Cox Field AAF
Cuero Field
Dalhart AAF
Dallas NAS
Dodd AAF
Dyess AFB
Eagle Pass AAF
El Paso AAF
Ellington JRB
Foster AFB
Galveston AAF
Galveston NAS
Garner Field
Gibbs AAF
Goodfellow AFB
Gray AFB
Harlingen AFB
Hensley Field (ANG)
Hicks Field (Camp Taliaferro)
Hitchcock NAS
Hondo AFB
Hulen AAF
James Connally ANGB
Jones Field
Kaufman "Erco" Field
Kelly Field
*Kingsville NAS
Lackland AFB
Lamesa AAF
Laredo AFB
Laughlin AFB
Louis Schreiner Field
Love Field
Majors Field AAF
Marfa AAF
Midland AAF
Mineral Wells AAF
Moore AFB
Pampa AAF
Pecos AAF
Penn Field
Perrin AFB
Pounds Field
Pyote AFB
*Randolph AFB
Reese AFB
Rich Field
San Angelo AAF
San Marcos AFB
*Sheppard AFB
South Plains AAF
Stinson Field
Terrell Field
Van Horn AAF
Webb AFB
Wink Airfield

*Denotes a contemporary, primary student flight-training facility.

2014. The USS *Lexington* now lives on as one of America's premier memorials, the "Museum on the Bay," in Corpus Christi, Texas. (Photo courtesy of the USS *Lexington*)

APPENDIX 2
A SAMPLING OF MILITARY AVIATION MUSEUMS IN THE GULF SOUTH

The Gulf Coast is home to numerous museums and monuments recognizing the heritage of military aviation. In addition, active military bases, as well as deactivated facilities, often feature displays, large and small, of aircraft and equipment. An all-inclusive listing of such displays is far too extensive for the purpose of this publication. For the most precise listing, consult a local visitors' bureau or military base public affairs office representing your geographic area of interest. Below are samplings of these extensive collections.

ALABAMA

Air Force Enlisted Heritage Hall, Maxwell-Gunter Air Force Base

Southern Museum of Flight, Birmingham

U.S. Army Aviation Museum, Fort Rucker

U.S. Space and Rocket Center, Huntsville

USS *Alabama* Battleship Memorial Park, Mobile

FLORIDA

Air Force Armament Museum, Eglin Air Force Base

Air Force Space & Missile Museum, Cape Canaveral Air Force Station

Armed Forces History Museum, Largo

Cpl. Larry E. Smedley National Vietnam War Museum, Orlando

DeLand Naval Air Station Museum, DeLand

Fantasy of Flight Museum and Collection, Polk City

Florida Air Museum, Lakeland

Florida Military Aviation Museum, Clearwater

Flying Tigers Warbird Restoration Museum/Kissimmee Air Museum, Kissimmee

Military Heritage Museum, Punta Gorda

National Naval Aviation Museum, Naval Air Station Pensacola

Valiant Air Command Warbird Museum, Titusville

GEORGIA

Marietta Museum of History, Marietta

Mighty Eighth Air Force Museum, Pooler

Museum of Aviation, Robins Air Force Base

LOUISIANA

Barksdale Global Power Museum, Barksdale Air Force Base

Chennault Aviation and Military Museum of Louisiana, Monroe

Jackson Barracks Military Museum, New Orleans

Louisiana Maneuvers and Military Museum, Pineville

Louisiana Military Museum, Ruston

National World War II Museum, New Orleans

Regional Military Museum, Houma

Southern Heritage Air Foundation, Tallulah

USS *Kidd* and Veterans Memorial, Baton Rouge

MISSISSIPPI

Memorial Training Aircraft Display, Naval Air Station Meridian

Mississippi Armed Forces Museum, Camp Shelby

Vietnam Veterans Memorial Project, Ocean Springs

TEXAS

1st Cavalry Division Museum, Fort Hood

12th Armored Division Memorial Museum, Abilene

American Airpower Heritage Museum, Midland

Battleship Texas State Historic Site, La Porte

Cavanaugh Flight Museum, Addison

Commemorative Air Force Airpower Museum, Midland

Flight of the Phoenix Museum, Gilmer

Fort Sam Houston Museum, Fort Sam Houston

Fort Worth Aviation Museum, Veterans Memorial Air Park Museum, Fort Worth

Freedom Museum USA, Pampa

Frontiers of Flight Museum, Dallas

H.E.A.R.T.S. Veterans Museum, Huntsville

Hangar-25 Air Museum, Big Spring

Lone Star Flight Museum, Galveston

National Museum of the Pacific War, Fredericksburg

National United States Armed Forces Museum, Houston

National Vietnam War Museum, Weatherford

National WASP World War II Museum, Sweetwater

Perrin Air Force Base Historical Museum, Denison

Silent Wings Museum, Lubbock

Southeast Texas War Memorial and Heritage Foundation, Orange

Texas Air Museum, Galveston

Texas Military Forces Museum, Austin

USS *Lexington*—Museum on the Bay, Corpus Christi

Vintage Flying Museum, Fort Worth

APPENDIX 3

GULF SOUTH MILITARY BASES AND STATIONS ON THE NATIONAL REGISTER OF HISTORIC PLACES

The U.S. National Park Service's National Register of Historic Places has recognized and is in the process of recognizing a number of facilities in the Gulf South for their contributions to military aviation heritage and history. The following sites are either on the National Register or in consideration for it:

ALABAMA

Maxwell Air Force Base, Montgomery

Tuskegee Airmen National Historic Site, Tuskegee

FLORIDA

Cape Canaveral Air Force Station, Patrick Air Force Base

Eglin Field Historic District, Fort Walton Beach

Glenn Curtiss House, Miami Springs

McKinley Climatic Laboratory, Eglin Air Force Base

Pensacola Naval Air Station Historic District, Pensacola

LOUISIANA

Barksdale Air Force Base Historic District, Bossier City

MISSISSIPPI

Terminal Building, Old Hangar and Powerhouse at Key Field, Home of the Mississippi Air National Guard 186th Refueling Wing and 185th Army Aviation Support Facility, Company G, Meridian

TEXAS

Hangar 9, Brooks Air Force Base, San Antonio

Kelly Field Historic District, San Antonio

Randolph Field Historic District, San Antonio

Stinson Field, Mission Parkway, San Antonio

USS *Lexington*, Corpus Christi

1950. The anchor once proudly displayed at Chevalier Field, NAS Pensacola. These historic grounds are now home to the Naval Education and Training Command. (Photo courtesy of the Emil Buehler Library, National Naval Aviation Museum, NAS Pensacola, FL)

ACKNOWLEDGMENTS

I wish to express my most sincere thanks and appreciation to each of the individual branches of the United States military for their research assistance with this book. A special thanks goes to the many public affairs office staff members and staff historians (including those named below and those unsung researchers behind the scenes) of the U.S. Air Force, U.S. Army, U.S. Coast Guard, U.S. Marine Corps, and U.S. Navy at the numerous bases, squadrons, and agencies who responded to my inquiries and requests.

Deep gratitude is extended to Dr. Robert Kane, U.S. Air Force historian, USAF Historical Research Agency, Maxwell Air Force Base, Alabama, for reviewing the final manuscript and offering additional historical and technical suggestions prior to publication. His expertise in Army and Air Force aviation history was an indispensable source of information for this book, as it is for anyone interested in military aviation research. Thank you, Dr. Kane!

In addition, I extend special thanks to Hill Goodspeed, acting director of the National Naval Aviation Museum, NAS Pensacola, Florida, for his review of the manuscript and assistance with finding rare photographs within the archives of the Emil Buehler Library at the museum.

Thank you to the following people and organizations:

Air Education and Training Command, Randolph AFB, JBSA

Air Force Historical Research Agency, USAF Air University, Maxwell AFB, Alabama

William Alexander

Annette Amerman, Historical Reference Branch, Marine Corps History Division

Capt. T. C. Balzer, Media Officer, USMC, Division of Public Affairs

Aileen Benavides, Eglin AFB

Caroline Brown

Etienne J. Caire II

Doug Campbell

Debbie Crites

Doris K. Crusoe, Columbus AFB

Amber Lynn Daniel, PAO, Blue Angels, USN

Jacqueline Davis

Mario Deluca, Historian, Blue Angels

Cherrie Dengler, USMC, JRB New Orleans

Richard Emmons, USAF

Tracy English

Kelly Fuller

Susan Griggs, USAF

Tasha Hairston, Randolph AFB

Kacy Hill, National World War II Museum, New Orleans

Ann Hussey, Randolph AFB

Maj. Paul Island, Deputy Director, OCPA, chief PAO

Jerry Jacob

Susan Junkins

Todd A. Kabalan, USAF

Don Koser

Robert Mitchell

Samuel O'Brian, USAF

Kelly Pate, U.S. Army

Andria Perez

Bruce Peterson

Lt. Col. Tom "Johnny" Ringo, USMC

Tim Savoir, USAF Armament Museum, Eglin AFB

Katherine Sowers, USCG

Lt. Matthew Stroup, Deputy Director, Navy Office of Information East

Third Combat Camera Squadron, Defense Imagery Management Operations Center (DIMOC), Lackland AFB, TX, producers of the video *100 Years of Aviation in San Antonio*

Denny Trice, NAS Key West

Dean Welch, U.S. Army

Harry White, NAS Pensacola

Cathy Whitney